Diachronic Perspectives on Domain-Specific English

Linguistic Insights

Studies in Language and Communication

Edited by Maurizio Gotti,
University of Bergamo

Volume 40

PETER LANG
Bern · Berlin · Bruxelles · Frankfurt am Main · New York · Oxford · Wien

Marina Dossena & Irma Taavitsainen (eds)

Diachronic Perspectives on Domain-Specific English

PETER LANG
Bern · Berlin · Bruxelles · Frankfurt am Main · New York · Oxford · Wien

Bibliographic information published by Die Deutsche Bibliothek
Die Deutsche Bibliothek lists this publication in the Deutsche Nationalbibliografie; detailed bibliographic data is available on the Internet at ‹http://dnb.ddb.de›.

British Library and Library of Congress Cataloguing-in-Publication Data: A catalogue record for this book is available from *The British Library,* Great Britain, and from *The Library of Congress,* USA

Published with a grant from Università degli Studi di Bergamo (Italy): Facoltà di Lingue e Letterature Straniere; Dip. di Lingue, Letterature e Culture Comparate.

ISSN 1424-8689
ISBN 3-03911-176-0
US-ISBN 0-8204-8391-5

Hochfeldstrasse 32, Postfach 746, CH-3000 Bern 9, Switzerland
info@peterlang.com, www.peterlang.com, www.peterlang.net

Contents

MARINA DOSSENA / IRMA TAAVITSAINEN

From Synchrony to Diachrony in Domain-specific English: An Introduction

In 2005 the 15th European Symposium on Language for Special Purposes ('New Trends in Specialized Discourse', held in Bergamo) hosted a workshop on the investigation of specialized discourse from a diachronic perspective. As a matter of fact, this workshop proved a very valuable opportunity for scholars of historical texts in specific branches of specialized discourse to present their findings and discuss their methodology in front of an audience of people who typically work on similar issues in present-day material and share the theoretical background. Studies of specialized discourse from a historical point of view often take modern linguistics as the point of departure. This feature was present in all papers of the workshop and consequently informs all the investigations in this volume. The first studies to trace the development of present-day conventions of English scientific and medical writing were published in the early 1990s (Atkinson 1992; Taavitsainen 1994). The relationship between contemporary linguistic research and the historical approach to scientific discourse has been outlined by Gotti (1996 and 2003), Atkinson (1999), and Taavitsainen (2001, 2002).

The research presented in this volume includes studies in the long-established tradition of scientific investigation of domain-specific English. The chapters offer new approaches, as cross-disciplinary points of departure are becoming increasingly common and cross-fertilisation is taking place between neighbouring fields like semiotics and lexical studies, as well as between more distant fields like legal history and discourse analysis. This volume also brings new data to the scholarly discussion, especially as far as new 'second-generation' historical corpora, either recently released or still in progress, are concerned. The coverage of domain-specific historical studies has certainly been improved by the availability of large one-genre one-register corpora designed according to state-of-the-art principles, and of task-specific small and medium-sized corpora. Transcriptions of authentic documents present another new trend in data acquisition. Returning to the original texts has made a crucial contribution to more reliable and focussed analyses of individual texts.

The methods of study in this volume range from rhetorical analysis and computer-aided qualitative analysis to statistically-based applications of corpus linguistics. Innovative uses are introduced, for example keyword analysis, and such studies may indicate future lines of development. Other innovations include the application of modern computer programs to trace the textual connections of the past. All texts have a historical anchoring, and diachrony also includes modern materials and limited time depth. Awareness of history can cast new light on material from recent decades, as such texts can be studied from the point of view of language change and development just as well as older texts. Scholars have started to wonder when texts become historical; for example, the first generation modern corpora Brown and London-Lund from the 1960s are already considered historical. Some contributions in this volume discuss recent trends of change in present-day domains, while the time depth of some other articles is much wider, ranging from Old English sources in seventeenth-century records to Late Medieval and Early Modern developments. Some of the studies are purely synchronic, focussing on one single text. Most contributions deal with English only, while some consider contact phenomena and the mutual intelligibility of different dialects and languages, or multilingual sources of vocabulary.

Discourse communities have become important anchorage points in modern LSP studies (Swales 1990; Barton 1994), but recent research has shown that the concept can be applied to past periods, with some modifications. Late Medieval discourse communities of specialised discourse are discussed by Jones (2004), and Early Modern by Taavitsainen / Pahta (1995) and Valle (1997 and 2003). Time- and place-specific sociocultural facts serve to provide explanations for manifestations of textual features in the past. This insight provides a central methodological feature of this volume, as all texts are studied in their sociohistorical frame. Modes of presentation and styles of writing change in the course of time in response to changing needs of the discourse community.

The contributions in this book deal with different fields of specialised discourse. There is increasing interest in the field of historical analysis of political rhetoric. Rhetorical analysis constitutes one trend in the field, while another takes its inspiration from modern discourse studies of critical linguistics with a focus on making underlying ideologies visible by the analysis of surface structures. In particular, investigations of historical news discourse have shed considerable light on this point (see Brownlees 2005).

Courtroom discourse has attracted increasing attention in recent scholarship and has become an important branch of historical pragmatics. Progress has been considerable, with new databases, more refined classification of materials, and innovative applications of pragmatic and corpus-linguistic approaches (see Archer 2005; Kryk-Kastovsky forthcoming). The language of law in general and the development of legal terminology have also received attention.

Medical and scientific writing has a long history in English. The tradition of learned writing can be traced to the fourteenth century, and the tradition of remedybook writing to Old English (Voigts 1984). Surgical treatises and texts with long traditions in learned Graeco-Roman writing started to be translated into vernacular languages in the late medieval period. Recent studies have shed light on the formation of generic and stylistic conventions as well as the multilingual practices of late medieval English (see the studies in Taavitsainen / Pahta 2004). Medicine was the spearhead of vernacularisation processes in English, and provides intriguing material now available to scholars on CD-ROM (Taavitsainen / Pahta / Mäkinen 2005).

Institutional and public discourse is central in most studies of historical developments in domain-specific English. The focus can be shifted to the private sphere as well, as specialised discourse can be studied in business correspondence. This becomes especially clear when we look at argumentative discourse, in which the formulation of the encoder's stance plays a crucial role in the attempt to persuade the recipient of the validity of the predication (Dossena 2001 and forthcoming). The construction of social networks in scientific exchanges follows similar criteria, and the principles seem to apply to both business and science networks and diplomatic exchanges (Dossena / Fitzmaurice 2006).

1. Organization and contents of the volume

The contributions in this volume could have been organized in various different ways: in chronological order, depending on the period they discuss; in thematic order, depending on the domain taken into consideration; or in methodological order, depending on the approach they take – whether they follow the tenets of corpus linguistics, or historical pragmatics, or historical lexicography. Our aim is to juxtapose studies in such a way as to highlight recent developments and foci of interest in the field. The book does not provide complete coverage, but reflects emerging trends and on-going research.

The first chapter investigates political discourse analysis with a rhetorical approach. NICHOLAS BROWNLEES studies argumentative strategies in pamphlets, a new and important medium of the Early Modern period used to distribute news and new ideas in religious and political spheres. Pamphlets were used to discuss controversial issues, and the rhetoric displays this purpose at the time of the Puritan Revolution. Brownlees' analysis of Gerrard Winstanley's rhetorical strategies highlights the encoder's virtual dialogue with the recipient (whether internal or external to the text itself), and his attempt to pre-empt criticism through awareness of possible objections. This was a common device in other fields of writing as well, e.g. in scientific texts.

Innovative methodology is one of the strengths of DAWN ARCHER's chapter. Building on statistically defined keywords that indicate the speaker's opinion (and therefore whether a question or a statement is more or less persuasive), it analyzes the ways in which defence and prosecution lawyers addressed witnesses, judges and juries before and after the Prisoners' Counsel Act (1836). The use of keywords is a major methodological innovation in corpus studies, and this is one of the first articles to discuss its implications for discourse studies (see also McEnery 2006). The author also offers methodological reflections on the study of historical texts, stressing the importance of acquiring complete transcriptions of original documents.

The investigation of the legal domain continues with contributions on lexical developments. The importance of sources for the compilation of a seventeenth-century legal glossary takes centre stage in PAOLA TORNAGHI's study of an unpublished manuscript by William Dugdale. Unlike typical 'hard word' dictionaries, this work appears to have a broader encyclopedic interest, aiming to illustrate the semantic value of individual items, but also including extensive discussion of the more ancient cultural worlds in which they were employed.

Still on legal discourse, HANS LANDQVIST looks at more recent diachrony, discussing changes that have occurred in Swedish legal terminology over the last fifty years owing to linguistic and social factors. Although this contribution is not directly about English, it is included here because it sheds light on the extent to which changing trends in international domains may affect individual languages. In such contexts, English has typically played a crucial role (see Cortese / Duszak 2005), and we thought it fruitful to broaden the scope to developments in other related languages in order to see how adaptations, calques and loanwords reflect cultural novelties in similar or different ways.

Contact phenomena are highlighted in the following two chapters. The historical roots of specialized lexis, this time in Older Scots administrative records, are discussed by JOANNA BUGAJ, whose investigation of binomials (in fact, 'binominals', as they are labelled in the analyses to which the author refers) presents the complex links existing in combinations of items of Germanic or Romance origin. Far from being mere translations of each other, they seem to have included additional shades of meaning or aimed at specific stylistic effects, as indicated by combinations of items deriving from the same language. The connection between Older Scots and other Germanic languages, particularly Scandinavian ones, is the object of MARJORIE LORVIK's analysis of vocabulary pertaining to the timber trade; as the author shows, the proximity of the languages employed by Scots and Scandinavian traders may actually have allowed them to conduct business effectively, without requiring the development of a *sui generis* pidgin.

Transcriptions of original documents are used as the material of study in the next two chapters, almost as a response to Archer's advocations. The first one, by GABRIELLA DEL LUNGO CAMICIOTTI, discusses the models of linguistic behaviour recommended in business correspondence style manuals. Basing her investigation on a specially-designed database of three late nineteenth-century manuals for Italian users, the author argues that stylistic changes in politeness-encoding strategies accurately reflect the changing status of the encoders; their constantly evolving interpersonal relations with their recipients allowed them to move from conventional deference to a more egalitarian standing. Indeed, the dynamic relationship between encoders and recipients in business settings is also discussed by MARINA DOSSENA, who presents findings from 19CSC (see Dossena 2004; Dury 2006). In particular, attention is given to the ways in which encoders introduce themselves and the purpose of their letter, with the assumption that this typically involves asymmetrical power relationships. The choice of positive and negative politeness moves is therefore dictated by a need to adopt 'fluctuating' identities related to the strength of the argument and, at the same time, how powerful an image the encoders wish to convey of themselves without posing threats to the recipient's face.

The following chapters rely on computerized materials, specifically the newly-released electronic corpus of *Middle English Medical Texts* and medical corpora still in progress. MARTTI MÄKINEN discusses important methodological issues for the identification of intertextuality in corpus material. Specific software is available for present-day documents to identify 'intertextuality', i.e. plagiarism in student essays, for example, but the situation becomes increasingly more complicated as we move back in time, when spelling conventions were still variable, and the very concept of textual adaptation, let alone plagiarism, was perceived in entirely different terms. The mere adoption of present-day tools thus becomes impossible, and new ones need to be developed, which may take into account the textual and cultural framework within which the texts were encoded.

The other two chapters on this domain focus on lexis, though different perspectives are offered. PÄIVI PAHTA's study analyzes the use of specific lexical items, i.e. amplifiers, for argumentative and other purposes in late medieval medical writing, thus highlighting the role of certain keywords in discourse. JUKKA TYRKKÖ discusses the semantic changes observed in the use of *sign*, *token*, *symptom* and *accident* in a corpus of medical writings that spans three centuries. The starting point is in modern semiotics and medical nosology. The gradual development of the meanings current in present-day usage is thus highlighted with numerous references to the sources employed by the documents under discussion, and to the scientific tradition they represented.

A more modern perspective on medical and legal writing is provided by ELENA SEOANE / CHRISTOPHER WILLIAMS. They compare recent changes in two apparently distant domains (the legal and the scientific) which can, however, allow us to identify interesting patterns of convergence in some respects, and divergence in others. Starting from the observation of the decrease in frequency of passive forms in scientific texts, and of *shall* in legal ones, the authors' main research question is whether we are witnessing a 'democratization' process in both domains. While this may seem so on the surface, findings show that – despite the recommendations of Plain Language pressure groups – legal texts still prove relatively more conservative, while changes in scientific ones attempt to convey the idea of dynamicity required by the present-day world even in academic contexts.

2. Future considerations

Considerable attention has been paid to methodological issues in all the contributions in this collection. The presentation of case studies would be insufficient, were it not supported by a clear theoretical approach, which can be translated into appropriate methods of investigation. As one of our referees put it, such case studies would be like "pictures of clouds" – maybe suggestive, but hardly meaningful from the scientific point of view if they are not suitably contextualized. This is all the more true when we look at documents pertaining to older stages of the language, for which suitable context may be difficult to retrieve. It is therefore all the more important that attempts to trace the history of domain-specific languages should be as wide-ranging as possible.

The collection shows that a great deal remains to be done in the field for a more comprehensive and clear picture of developments. The broad outline has already become visible, but there are still gaps in our knowledge and more studies are needed. We hope that this volume will provide inspiration and further incentives to pursue studies in domain-specific English.

Acknowledgements

The editors would like to thank the organizers, participants, and discussants of the above-mentioned Workshop for their contribution to the debate. We are also grateful to Maurizio Gotti (editor of the *Linguistic Insights* series) and the anonymous referees of the volume for their comments on individual papers and on the edited collection. Any remaining faults in the latter are, of course, the responsibility of the editors.

Bergamo / Helsinki, May 2006

References

Archer, Dawn 2005. *Questions and Answers in the English Courtroom (1640-1760): A Sociopragmatic Analysis*. Amsterdam: Benjamins.

Atkinson, Dwight 1992. The Evolution of Medical Research Writing from 1735 to 1985: The Case of the *Edinburgh Medical Journal*. *Applied Linguistics* 13, 337-374.

Atkinson, Dwight 1999. *Scientific Discourse in Sociohistorical Context:* The Philosophical Transactions *of the Royal Society of London, 1675-1975*. Mahwah, NJ: Lawrence Erlbaum.

Barton, David 1994. *Literacy: An Introduction to the Ecology of Written Language*. Oxford: Blackwell.

Brownlees, Nicholas (ed.) 2006. *News Discourse in Early Modern Britain*. Bern: Peter Lang.

Cortese, Giuseppina / Duszak, Anna (eds) 2005. *Identity, Community, Discourse. English in Intercultural Settings*. Bern: Peter Lang.

Dossena, Marina 2001. "For the aduancement of the commoun weilth": Hypotheses in Scots Scientific Discourse of the Sixteenth-Seventeenth Centuries. A Preliminary Investigation. *Scottish Language* 20, 47-65.

Dossena, Marina 2004. Towards a Corpus of Nineteenth-century Scottish Correspondence. *Linguistica e Filologia* 18, 195-214.

Dossena, Marina forthcoming. Doing Business in Nineteenth-century Scotland: Expressing Authority, Conveying Stance. *IEEE Transactions on Professional Communication – Insights from Corpus Linguistics for Professional Communication.* Guest editor: Thomas Orr, University of Aizu (Japan).

Dossena, Marina / Fitzmaurice, Susan M. (eds) 2006. *Business and Official Correspondence: Historical Investigations*. Bern: Peter Lang.

Dury, Richard 2006. A Corpus of Nineteenth-century Business Correspondence: Methodology of Transcription. In Dossena / Fitzmaurice (eds), 193-205.

Gotti, Maurizio 1996. *Robert Boyle and the Language of Science*. Milano: Guerini.

Gotti, Maurizio 2003. *Specialized Discourse: Linguistic Features and Changing Conventions*. Bern: Peter Lang.

Jones, Claire 2004. Discourse Communities and Medical Texts. In Taavitsainen / Pahta (eds), 23-36.

Kryk-Kastovsky, Barbara (ed.) forthcoming. *Historical Court Room Discourse. Journal of Historical Pragmatics,* Special issue, 7/2.

McEnery, Tony 2006. The Moral Panic about Bad Language in England, 1691-1745. *Journal of Historical Pragmatics* 7/1, 89-113.

Swales, John M. 1990. *Genre Analysis: English in Academic and Research Settings*. Cambridge: Cambridge University Press.

Taavitsainen, Irma 1994. On the Evolution of Scientific Writings from 1375 to 1675: Repertoire of Emotive Features. In Fernández, Francisco / Fuster, Miguel / Calvo, Juan José (eds) *English Historical Linguistics 1992*. Amsterdam: Benjamins, 329-342.

Taavitsainen, Irma 2001. Evidentiality and Scientific Thought-Styles: English Medical Writing in Late Middle English and Early Modern English. In Gotti, Maurizio / Dossena, Marina (eds) *Modality in Specialized Texts*. Bern: Peter Lang, 21-52.

Taavitsainen, Irma 2002. Historical Discourse Analysis: Scientific Language and Changing Thought-styles. In Fanego, Teresa / Méndez-Naya, Belén / Seoane, Elena (eds) *Sounds, Words, Texts, Change: Selected Papers from the Eleventh International Conference on English Historical Linguistics*. Amsterdam: Benjamins, 201-226.

Taavitsainen, Irma / Pahta, Päivi 1995. Scientific 'Thought-styles' in Discourse Structure: Changing Patterns in a Historical Perspective. In Wårvik, Brita / Tanskanen, Sanna-Kaisa / Hiltunen, Risto (eds) *Organization in Discourse: Proceedings from the Turku Conference*. Turku: Univ. of Turku, 519-529.

Taavitsainen, Irma / Pahta, Päivi (eds) 2004. *Medical and Scientific Writing in Late Medieval English*. Cambridge: Cambridge University Press.

Taavitsainen, Irma / Pahta, Päivi / Mäkinen, Martti (eds) 2005. *Middle English Medical Texts*. Amsterdam: Benjamins.

Valle, Ellen 1997. A Scientific Community and its Texts: A Historical Discourse Study. In Gunnarsson, Britt-Louise / Linell, Per /

Nordberg, Bengt 1997. *The Construction of Professional Discourse.* London: Longman, 76-98.

Valle, Ellen 2003. "Let me not lose yr love & friendship". The Negotiation of Priority and the Construction of a Scientific Identity in Seventeenth-century Natural History. In Hiltunen, Risto / Skaffari, Janne (eds) *Discourse Perspectives on English: Medieval to Modern.* Amsterdam: Benjamins, 197-234.

Voigts, Linda Ehrsam 1984. Medical Prose. In Edwards, A.S.G. / Pearsall, Derek (eds) *Middle English Prose: Essays on Bibliographic Problems.* New York: Garland, 315-335.

NICHOLAS BROWNLEES

Gerrard Winstanley and Radical Political Discourse in Cromwellian England

1. Introduction

In support of his view that "modern political thought begins in the English revolution of the seventeenth century", Hill (1983: 9) refers to three key political writers of that era. They are Thomas Hobbes, James Harrington and Gerrard Winstanley. Thomas Hobbes's *Leviathan* and James Harrington's *Oceana* were published in 1651 and 1656 respectively, while Gerrard Winstanley's writings spanned the four years from 1648 to 1652.

In this study I shall be examining Winstanley's use of spoken discourse in his final, most ambitious political tract *The Law of Freedom in a Platform* (1652). This pamphlet of 38,000 words was published two years after the collapse of the agricultural commune that he had helped set up on common land in Surrey in 1649 with other so-called Diggers. The Diggers were a small but significant group of radical Protestants whose primary political belief was the common ownership of property. Although Winstanley's practical experiment in prototypical communism was unsuccessful, this did not prevent him from determining to set out his blueprint, or "platform", for a radically different society. What Winstanley envisages in *The Law of Freedom* is the common ownership of land, the abolition of the buying and selling of land and labour, the communal practice of agriculture, representative republican government, a totally different legal system with no professional lawyers, education for both men and women, and total religious toleration. Only in such a way could the new Commonwealth "finde out where true Freedom lies" (*LF* 1652:

17).[1] While this utopian vision of mankind's potential had no visible effect on Oliver Cromwell, England's *de facto* ruler in 1652, it has been studied and approvingly quoted in radical socialist and communist literature over the last 100 years. *The Law of Freedom* is recognised as being one of the most significant political works of seventeenth-century England.

In my analysis of *The Law of Freedom* I shall focus on the functions and forms of monologic and dialogic spoken discourse in the work. In considering functional motivation, I shall examine why voice is given to particular personages and groups, and how the framing argumentative strategies in which the speech is formulated aid Winstanley's political object. Attention will be given to the different kinds of animadversion adopted by Winstanley, and how one exemplar of animadversion, procatalepsis, is particularly successful in furthering the author's aim. As to form, I shall examine the linguistic representation of Winstanley's use of orality, and the manner in which this conforms to, or differs, from conventions at that time. In assessing how far the fictitious conversations employed by the author approximate natural spoken language, reference will be made to Biber's analysis of spoken interaction, where one key dimension distinguishing oral from literate discourse is that of "involved versus informational production" (Biber 1988: 104-108). However, before considering Winstanley's use of spoken discourse, reference must be made to the role and textual characteristics of the pamphlet in the English Revolution.[2] This contextualisation is essential if one is to understand why and how Winstanley exploits orality in his political tract.

---

1 I shall use the abbreviation *LF* in bibliographical references to the *Law of Freedom*. The text I refer to is part of the Thomason Collection at the British Library, London. As the edited versions in Sabine (1965) and Hill (1983) have standardised spelling, they have not been adopted for citation purposes.

2 In adopting the term 'English Revolution' I am referring to the dramatic military, political and social events that occurred in England between 1640 and 1660.

2. The pamphlet

The role of pamphlet publications in mid-seventeenth century England cannot be underestimated. From the breakdown of royal government and censorship in 1641, printed publications, and particularly those in pamphlet form, provided a forum for a hitherto unimaginably large number of writers and opinions.[3] Never before had so many unknown, unheard Englishmen been able to express their often wildly contrasting ideologies. The cheap pamphlet was the quintessential means of communication for the publication of news, religious and political polemic, and, as in Winstanley's case, the dissemination of a radical political manifesto.

The pamphlet not only provided access to a host of new voices, it also reached out to an increasingly large cross-section of society. As "England had an unusually high level of popular literacy" (Zaret 2000: 174) political pamphlets in the 1640s were not just read by the educated elite but also by less sophisticated readers. These readers, like some of the authors whose works they read, or heard read out aloud, often had little formal education and were generally unversed in the intricacies of classical rhetoric. What revolutionary England had created was a new author and a new audience, both of whom were influenced by an oral culture characteristic of those of little formal education.[4]

The fact that the pamphlet created a public sphere in which the topical issues of the day could be discussed, and public opinion shaped, was in no small measure the result of the inherent textual characteristics of the medium itself.[5] The overriding trait of the mid-

3 For wide-ranging analyses of pamphlet authors, contents, printing practices and readership in Early Modern Britain see Raymond (2003: 53-97) and Zaret (2000: 133-216).

4 See Fox (2000) for a fascinating account of the interrelationship between spoken and written forms of communication in Early Modern Britain.

5 See Zaret (2000: 21-34) for a critique of how the original concept of the public sphere, as set out in Habermas (1962), has been revised and developed. Along with other contemporary historians, in particular Raymond (1999), Zaret argues that in England the creation of the public sphere can be traced back to the 1640s.

seventeenth-century pamphlet was its textual heterogeneity. According to Raymond, "the heterogeneity of the pamphlet form enabled pamphleteers to compress news, history and opinion into a few sheets and to bring into focus diverse heterogeneous materials and voices" (2003: 214). The multifaceted nature of pamphlet texts meant that in any single publication one could find a strikingly high variety of text types.[6] The variety of discourse models both reflected and accommodated not only the widely diverse nature of pamphlet writers and their topics but also the frequently variegated nature of pamphlet readership.

Such heterogeneity of content and discourse modes meant pamphlets were often capable of appealing to various levels of the reading public. For example, in his 1648 political tract about the Independents,[7] Clement Walker recognised his own intentional diversity of style, saying he had written "in a mixed style, in which [...] there are some things to hold the judgments of the gravest; some things fit to catch the fancies of the lightest, and some things of a middle nature, applying myself to all capacities" (Zaret 2000: 199). Regarding such heterogeneity *The Law of Freedom* is no exception. Although the pamphlet is unequivocally political in its intention and contents, it nevertheless includes not just closely reasoned formal political argumentation, but also verse, two epistles, legal commentary and numerous instances of spoken discourse.

---

6 I follow Taavitsainen in distinguishing genre from text type. Regarding the former Taavitsainen says: "Genres are inherently dynamic cultural schemata used to organise knowledge and experience through language" (2001: 139-140). Text types, on the other hand, are classified on the basis of text-internal features.

7 In the 1640s the more radical and extreme opponents of the king were frequently referred to as Independents.

3. Spoken discourse in *The Law of Freedom*

3.1. The epistle dedicatory

The Law of Freedom is divided into three parts. The first two sections of the treatise consist of separate epistolary addresses to Oliver Cromwell and the reader. The final section contains the platform's full exposition. In my analysis I shall concentrate on the epistolary address to Cromwell and the platform, or manifesto, itself. The epistle addressed to the reader will be omitted since its short text contains no spoken discourse.

The first epistle, or "epistle dedicatory" as it is called by Winstanley, is 5,430 words long and respects epistolary conventions in so far as it begins with the salutation 'sir', and finishes with the suitably humble formula: "And so I leave this in your hands, humbly prostrating myself and it before you, and remain [...]" (*LF* 1652: 14). However, while epistolary in form, this address to Cromwell should be considered as political in its intention and content as the platform itself. As regards content, the epistle dedicatory provides a justification of the most significant points in the programme, and as for intention one must bear in mind Winstanley's words in the pamphlet *A Watch-Word to the City of London, and the Armie* (1649: preface): "thoughts run in me that words and writings were all nothing, and must die, for action is the life of all, and if thou dost not act thou dost nothing". Since Winstanley was no abstract theorist, but wrote to achieve a purpose, his epistle dedicatory had a specific political objective too. In the epistle he is referring to and justifying the main points of the Diggers' manifesto in the hope that Cromwell would implement it.[8]

Within the dedication one finds Winstanley not only addressing Cromwell but also establishing a dialogic relationship with him. This has immediate symbolic importance in that it indicates Winstanley's recognition that he, and other fellow Diggers, has the right to engage

8 See Corns (1992: 161): "Winstanley's Digger pamphlets do not speak primarily to the propertiless: rather, they speak for the poor oppressed people of England to the great ones of the earth."

in political discussion with the most powerful man in the land. This exchange is primarily achieved through different types of animadversion including, in particular, procatalepsis. Procatalepsis was a well-known rhetorical strategy which consisted of the writer anticipating and answering an opponent's objections. In Winstanley's epistle it is Cromwell who frequently raises objections to the Diggers' proposals and Winstanley who provides clarification. Cromwell's comments are usually preceded by the deictic 'you' and introductory reporting verb. When these linguistic features are absent one finds free direct speech where Cromwell is still the supposed speaker.

(1) You will say, We must be subject to the Ruler, it is true, but not to suffer the Rulers to call the earth theirs and not ours […]. (*LF* 1652: 9)

(2) Is not buying and selling a religious Law? No, it is the law of the Conqueror, but not the righteous Law of Creation (*LF* 1652: 12)

(3) But shall not one man be richer then another?
There is no need of that; for Riches make men vain-glorious, proud, and to oppress their Brethren […]. (*LF* 1652: 12)

As (1-3) indicate, Cromwell's objection may have either declarative or interrogative form. The use of the interrogative is not surprising since as Claridge writes: "questions as a classic tool in persuasive and polemical speech have an important role to play in pamphlet literature" (2005: 133).

The procatalepsis sometimes consists of more than one isolated objection and its successive rebuttal.

(4) It may be you will say to me, *What shall I do*? I answer, You are in place and power to see all Burthens taken off from your friends, the *Commoners of England*. You will say, *What are those Burthens*?
I will instance in some […]. (*LF* 1652: 5)

Towards the conclusion of the epistle we find a very emphatic rebuttal of Cromwell's anticipated objections.

(5) If you say, Some will live idle; I answer, No: It will make idle persons to become workers, as is declared in the *Platform*; There shall be neither Beggar nor idle person.
If you say, This will make men quarrel and fight:

> I answer, No: It will turn swords into plowshares, and settle such a peace in the Earth, as Nations shall learn War no more. (*LF* 1652: 14)

The force of Winstanley's counterargument is provided by the repetition of the emphatic "I answer, No", accompanied by explanatory argumentation, which in the second instance above is particularly detailed. What is predictable, but not for that insignificant, is that throughout Winstanley's fictitious dialogues with Cromwell the exchange is so structured that the Digger's responses are conclusive, allowing no further space for discussion. Procatalepsis, like other forms of animadversion, is an effective rhetorical device in that while appearing dialogic, in practice it allows the author to remain in complete control of the discourse. With such forms of animadversion "readers might be left with an impression that they were onlookers to an evenly contested dialectic but in fact there could be only one winner in the debate" (Brownlees 2006: 21). What we see in Winstanley's dialogic exchange with Cromwell is that although the former is speaking to England's political and military leader that does not prevent him from taking full advantage of the rhetorical possibilities offered by procatalepsis.[9]

However, the oral element in the epistle dedicatory is not confined to Winstanley's exchange with Cromwell. The other participants in dialogic interaction include 'the people', 'soldiers', 'lawyers', 'men of other Nations' and Winstanley's own 'spirit'. The participants who speak most often are the people. We frequently hear their voice through both direct and indirect speech. In introducing the voice of England, Winstanley makes it clear that the addressee of the people's social and political grievances is Cromwell:

(6) I hear the people dayly complaining of, and groaning under, looking upon you and waiting for Deliverance.
Most people cry, We have payd Taxes, given Freequarter, wasted our Estates, and lost our Friends in the Wars, and the Taskmasters multiply over us, more than formerly. (*LF* 1652: 5)

---

9 Winstanley's lack of deference towards Cromwell is also indicated by his use of deontic modality when giving advice to England's leader: "And now you have the Power of the Land in your hand, you must do one of these two things [...] Or secondly, you must onely remove the Conquerors Power out of the Kings hand into other mens [...]" (*LF* 1652: 4).

The inclusion of the people's voice is functionally very important. By apparently citing their actual words, Winstanley is showing Cromwell that the Diggers' programme is not the elaboration of a small sect but the expression of an entire people – the people of the English Commonwealth.

(7) For say the people, If the Lords of Manors and our Task-masters hold Title to the Earth over us from the old Kingly power, behold that power is beaten and cast out. (*LF* 1652: 8)

(8) Now saith the people, By what Power do these maintain their Title over us? Formerly they held Title from the King, as he was the Conquerors successor: But have not the Commoners cast out the King, and broke the bond of that Conquest? (*LF* 1652: 7)

On several occasions Winstanley also engages with the people. The Digger places himself as the intermediary between the people and their leader.

(9) I have asked divers this question, *Why do you say so?*
Some have answered me, that Promises, Oaths, and Engagements have been made as a motive to draw us to assist in the Wars [...] (*LF* 1652: 6-7)

Given that the Diggers' readership included both active and disbanded soldiers it is no surprise that in one instance we hear their voice too:

(10) I have asked divers Souldiers what they fought for; they answered, they could not tell; and it is very true, they cannot tell indeed, if the Monarchical Law be established without Reformation. (*LF* 1652: 7)

Likewise, on one occasion, we are provided with the direct speech of lawyers. The passage occurs when Winstanley is explaining the population's anger at legal malpractice. As with the people's and soldiers' words, the presence of the lawyers' speech aims to authenticate Winstanley's argument by providing the reader with direct access to the protagonists' thinking. The reader is not just acquainted with Winstanley's interpretation of their views but with what the various protagonists actually say.

(11) So that we see, though other men be under a sharp Law, yet many of the great lawyers are not, but still do act their will, as the Conqueror did; as I have heard some belonging to the Law say, *What cannot we do?* (*LF* 1652: 7)

The last voice to mention is Winstanley's spirit. This participant is mentioned just once, near the beginning of the epistle, and the author employs the spirit's words to explain to Cromwell why he is addressing him so straightforwardly. Winstanley's explanation is that were his message to be anything but clear and direct he could risk Cromwell not fully understanding the full import and significance of the manifesto.

(12) Sir, I pray bear with me; my spirit is upon such a lock that I must speak plain to you, lest it tell me another day, If thou hadst spoke plain, things might have been amended. (*LF* 1652: 5)

In (12) the important word is 'plain'. Since the publication and controversy of the first Marprelate tracts in 1588, 'plain-speaking' had been a recognised virtue of Protestant writing (Raymond 2003: 27-52). In writing 'plainly' Protestant authors were consciously distancing themselves from the supposedly ornate and mendacious discourse of their adversaries. Most illustrative of this desire for plain-speaking is Marprelate's declaration that he "must needs call a Spade a Spade, a Pope a Pope" (1588: sig. A2r). What we need to decide in our analysis of the *Law of Freedom* is to what extent this desire for plain speaking also influences Winstanley's representation of spoken discourse.

3.1.1. Representation of orality in the epistle dedicatory

Analysis of the spoken discourse in the epistle dedicatory indicates that with the exception of the spirit's utterance, where we find the informal 'thou' pronoun, and the occasional use of 'yes/no' response words, there is little linguistic difference between spoken passages and the rest of the text in this section.[10] While the spoken discourse expresses characteristic features of plain-speaking such as clear

10 Although there is not always a clear, textually identifiable boundary between what is spoken and non-spoken, the former undoubtedly amounts to more than half of the epistle's contents.

argumentation, straightforward syntax, and a mixture of general and specific vocabulary, as do the other parts of the epistle, it does not include features which Biber, in his analysis of the interactive dimension of communication, considers typically dialogic. In this analysis, where dialogic interaction is equated with "involved" rather than "informational production" (Biber 1988: 104-108), characteristic features of involved communication include "a reduced surface form, a generalized or uncertain presentation of information, and a generally fragmented production of text" (Biber 1988: 106). These features, which are typically represented by contractions, pronominal forms, indefinite pronouns, final (stranded) prepositions, pro-verb *do*, hedges, lexical repetition and imprecision, are generally absent in Winstanley's fictitious conversations.[11] The following paragraphs, containing both direct speech and literary argumentation, exemplify the overall homogeneity of the language in the epistle. Extract (13) begins with Winstanley addressing Cromwell, continues with Cromwell's anticipated objection and finishes with Winstanley's counter-argument. In (14) we first find Winstanley's comment on a particular grievance, followed by the supposed words of the people.

(13) Though this Platform be like a peece of Timber rough hewd, yet the discreet workmen may take it, and frame a handsome building out of it.
It is like a poor man that comes cloathed to your door in a torn country garment, who is unacquainted with the learned Citizens unsetled forms and fashions; take off the clownish language, for under that you may see beauty.
It may be you will say, If Tythes be taken from the Priests and Impropriators, and Copy-hold Services from Lords of Mannors, how shal they be provided for again; for is it not unrighteous to take their estates from them?
I answer, when Tythes were first enacted, and Lordly power drawn over the backs of the oppressed, the Kings and Conquerors made no scruple of Conscience to take it, though the people lived in sore bondage of poverty for want of it [...]. (*LF* 1652: 11)

(14) Fourthly, There is another grievance which the people are much troubled at, and that is this; Country people cannot sell any Corn or other fruits of the Earth in a Market Town, but they must either pay Toll, or be turned out of Town: now say they, This is a most shameful thing, that we must part with our

11 See also Jucker (2004) for general methodological issues in historical dialogue analysis and Taavitsainen (1999) for an examination of orality, and how it can be defined, in Early Modern medical dialogues.

> estates in Taxes and Free-quarter to purchase the Freedom of the Land, and the Freedom of the Towns, and yet this Freedom must be still given from us, into the hands of a covetous Norman Toll-Taker, according to the Kings old burdensome Laws, and contrary to the Liberty of a free Commonwealth.
> (*LF* 1652: 8)

Although (13) and (14) contain five separate communicative contexts, there seems to be no variation in linguistic features to reflect such variety. The spoken discourse in the people's and Cromwell's words, and Winstanley's answer to Cromwell, are entirely similar to Winstanley's address to Cromwell as well as his general comment in (14) where he has no addressee in mind. We therefore find Winstanley using spoken discourse in the epistle but investing such discourse with very few specific linguistic features common to spoken language. To explain the reason for this, it is necessary to consider both the general representation of direct speech in mid-seventeenth-century pamphlets and Winstanley's own rhetorical reasons for not linguistically marking the spoken discourse of his protagonists in the epistle.

Direct speech is intermittently found in a wide range of mid-seventeenth-century pamphlets. For example, in the news pamphlets of the day we occasionally find direct speech in narrative news reports, and in particular those where the news or more frequently the propaganda is expressed in dialogue form, as well as in the reporting of parliamentary proceedings, speeches of political personages, religious address and liturgy, court proceedings, and the public confessions of criminals and other assorted ill-doers. However, although writers make use of direct speech, they very infrequently mark it with some of the characteristics associated with such discourse. For example, the spoken discourse of political personages is invariably more reflective of the literate than the oral style (Brownlees 2005: 74-81).[12] Foregrounding in the textual re-

---

12 Typical of such discourse is Archbishop Laud's scaffold speech in January 1645. His final public address begins: "*Good people*; I am an old man, and my memorie short, and therefore being come to this place, I must crave leave to make use of my papers: I dare not trust to my memorie" (*A Perfect Diurnall of Some Passages in Parliament* (1645: 603). This kind of set speech is often moving in its expression of loss, regret, or suffering, but linguistically it contains few of the characteristics indicated by Biber as being typical of interactive communication (Brownlees 2005: 76).

presentation of spoken discourse only usually occurs where the speakers are either intentionally aiming at humour, as in political satire, or are the object of comedy or social disapproval or both.[13]

Given how spoken discourse is represented in mid-seventeenth-century pamphlets it is not surprising that Winstanley refrained from foregrounding spoken discourse features in his epistle dedicatory. Since Cromwell was a political personage worthy of the utmost respect and dignity, his voice had to reflect the conventions of standard written English. Anything else would have been marked and probably interpreted as a criticism of Cromwell. Likewise, the fact that the other main participants in the epistle, Winstanley and the people, also speak in a literate tone of voice can be explained by reference to extract (13). Here Winstanley commends the platform to the attention of England's leader even if the language in which the document is expressed lacks decorum and civility. Cromwell is asked to "take off the clownish language, for under that you may see beauty" (*LF* 1652: 11). These words indicate Winstanley's keen understanding of the close relationship between form and content. In order to achieve as forceful a political message as possible, he had to assure Cromwell and his other readers that he, the people, and the soldiers (for whom he also had sympathy) "were not clownish", but deserved to be heard. This might not have been the case if Winstanley had invested his own and the people's spoken discourse with marked oral forms.[14] The only group of participants who might well have been given marked oral forms, as an expression of Winstanley's disapproval, were the lawyers. The fact that their speech, which is limited to the one self-damning utterance in (11), is presented in standard literate English

---

13 See Brownlees (2006), McElligott (2004) and Raymond (1996: 155-158) for an analysis of orality and political satire in English Civil War newsbooks. For an example of spoken discourse in bawdy pamphlet comedy, see *The Wandring Whore* (1660: 2, 7-8). Social satire is instead found in *The Welch Mercury* (21-28 October 1643), where the spoof Welsh publication contains numerous nonstandard spoken discourse features, representative of the Welsh people's ignorance in general.

14 In describing the low esteem in which common people were held in Early Modern Britain, Fox writes: "the 'vulgar', it was always said, were credulous and gullible in all that they heard, ever liable to misunderstand the truth of things, prone to distort them still further, hasty to judge, and quick to criticize their betters" (2000: 339).

does not however indicate that Winstanley was unaware of the possibilities of negative characterization through marked orality. In the following analysis of the platform, the third and last part of *The Law of Freedom*, we see Winstanley making specific, albeit limited use of foregrounded orality, as an expression of authorial censure.

3.2. The Law of Freedom in a platform

The platform is divided into six chapters dealing with separate points in the manifesto. Although spoken discourse is less frequent in this section than in the epistle dedicatory, it still achieves similar functional purposes. It establishes authorial authenticity, provides insight into the character and moral rectitude of the speaker, and gives dramatic voice. Regarding this last point, one must not forget that although the work is addressed to Cromwell, and therefore implicitly to other protagonists in contemporary English society, the pamphlet's readership would also have consisted of quite humble people. As regards the readership of Digger pamphlets, Smith writes that the bulk of sectarian membership, which included the Diggers, consisted of "the lower professions and merchants, the artisanal groups in the cities, and the yeomanry in the country" (1989: 11). Such people were probably educated – though like Winstanley himself not necessarily the beneficiaries of a formal education – but without great expertise or sophistication in the appreciation of written polemic.[15] Thus, Winstanley's recourse to the presentation of ideas through spoken discourse and dialogic interaction not only gives the manifesto a dramatic element that would have been appreciated by some readers, but also facilitates the comprehension of the political message. This last aspect is supported by James Harrington, who justified his own use of dialogue in the political tract *Valerius and Publicola* (1659) by saying it was: "the clearest and most effectual for conveying a mans sense into the understanding of the Reader" (Toland 1700: 475).[16]

---

15 For biographical information on Winstanley, see Alsop (2000).

16 As Taavitsainen / Nevanlinna (1999) illustrate, the didactic function of dialogue was also appreciated by some sixteenth-century authors.

Although monologic speech is found in the platform, in the main spoken discourse is most often expressed by animadversion.[17] Here, however, the animadversion does not only include anticipated objections but also those that for argumentative purposes have already been formulated. Winstanley's interlocutors include Cromwell and another group vaguely referred to as "some". Unlike the 'people', who support the Diggers' proposals, the 'some' group put forward the traditional viewpoint.

(15) But you will say, If it must be so, then will men so differ in their judgments, that we shall never agree, I answer:
There is but Bondage and Freedom, particular interest [...] (*LF* 1652: 53)

(16) Some say, It lies in the free use of Trading, and to have all Pattents, Licenses, and Restraints removed: But this is a *Freedom* under the Will of a Conqueror. Others say, It is true Freedom to have Ministers to preach, and for people to hear whom they will, without being restrained or compelled from or to any form of worship: But this is an unsetled Freedom. (*LF* 1652: 17)

Similarly adversarial are ecclesiastical 'ministers' (*LF* 1652: 21), and a 'zealous, but ignorant professor'. In (17), the latter's objection to Winstanley's statement is preceded by the discourse marker "I, but" (*LF* 1652: 58), *I* being an EModE variant of the response word *Ay(e)*:[18]

(17) By this means, in time men shall attain to the practical knowledge of God truly; That they may serve him in spirit and truth; and this knowledge will not deceive a man.
I, but saith the zealous, but ignorant Professor,
This is a low and carnal ministry indeed, this leads men to know nothing, but the knowledge of the earth and the secrets of nature, but we are to look after spiritual and heavenly things. (*LF* 1652: 58)

The use of "I, but" is also found in an extended dialogue between an 'elder and younger brother'. In Digger ideology the elder brother

---

17 Monologic speech is seen when the 'people' address their 'officers' (*LF* 1652: 35), a child addresses her father (*LF* 1652: 34), and a wrongdoer asserts his right to private property (*LF* 1652: 43).

18 In her analysis of two EModE trials, Kryk-Kastovsky (2000: 16) notes the relatively high frequency of *Ay* as a discourse marker.

portrayed doctrinaire reason and the suppression of true freedom while the younger brother personified the attempt to supplant the cold hand of tradition with fresh life and liberty.

(18) I, but saith the elder Brother, *You must not trust to your own Reason and Understanding, but you must believe what is written and what is told you* [...] (*LF* 1652: 61)

It is not possible to know if the discourse marker 'I, but' belonged to a specific register, but even if it did it would be mistaken to associate it with authorial disapproval of the speaker since on one occasion Winstanley himself makes use of the expression.[19] The only instance in the platform where we find the negative contents of a protagonist's message seemingly reinforced by linguistic foregrounding in the spoken discourse is in the case of the "spirit of Monarchy" (*LF* 1652: 27). This spirit, which is also referred to as "this subtle spirit of darkness" (*LF* 1652: 28), is described in some detail as 'he' who insidiously seduces the 'people' into subjugation:

(19) For in that man wherein this kingly spirit seats himself, he tells the people that are wronged, *Well, I'le ease you, and I'le set things to rights*: And then he went about to establish buying and selling by Law, whereby the people had some ease for a time, but the cunning Machavilian spirit got strength thereby to settle himself King of the Earth.

For after some time the people through ignorance began to multiply Suits of Law one against another, and to quarrel and fight: Now saith this subtle spirit, *Come follow me*, to one sort of people that are oppressed, *and stick to me, and we will fight with those who wrong you; and if we conquer them, then we will govern the Earth as we please, and they shall be our servants, and we will make them work with us*. (*LF* 1652: 28)

The kingly spirit's first utterance (*Well, I'le ease you, and I'le set things to rights*) is the only example of spoken discourse in the entire work which unmistakeably reflects colloquial usage. The sentence includes the discourse marker "Well", the only examples of contracted verbal forms in spoken discourse throughout the entire tract, and the

19 When rebutting "the National Clergy of England and Scotland", Winstanley says: "I, but this is but the cry of an Egyptian Task-master [...]". (*LF* 1652: 21).

idiomatic expression '*set things to right*'.[20] Such a marked conjunction of orality must reflect specific authorial design: Winstanley aims to underline the insidiousness of the kingly spirit's message, and the spirit appears to adopt discursive, demotic forms so as to deceive and eventually subjugate the British people.[21]

Although this marked orality is only found in the spirit's first speech, it would be wrong to think its characteristic nature is limited to that utterance alone. In the reader's mind the explicit oral quality of the first speech carries over to the rest of the discourse.[22] Winstanley provides enough orality in the spirit's discourse to indicate duplicity, but not too much to transform such perversity of intent into comedy. Had the spirit continued using the contracted forms and other marked oral traits in its other utterances it would have lost its menace and just appeared as a figure of fun.

To complete this analysis of Winstanley's use of oral markedness, we can refer to another passage, in which the author refers metalinguistically to the purity of language:

(20) And thus to speak, or thus to read the Law of nature (or God) as he hath written his name in every body, is to speak a pure language, and this is to speak the truth as Jesus Christ spake it [...] (*LF* 1652: 58)

Bearing in mind how spoken discourse is represented in the *Law of Freedom in a Platform* one must conclude that for Winstanley language purity was a feature of literate rather than oral discourse.

---

20 In referring to the oral characteristic of discourse markers such as 'well', Biber *et al.* write: "Discourse markers are particularly characteristic of spoken dialogue. They are words and expressions which are loosely attached to the clause and facilitate the ongoing interaction" (1999: 140). For the use of *well* and other hedges in early English dialogues see Jucker (1997) and Culpeper / Kytö (1999).

21 The exploitation of textual orality has been interestingly analysed by critical linguists such as Fowler (1991) in relation to the creation of modern-day ideological consensus between British newspapers and their readers.

22 See Page (1988: 55-121) for how novelists succeed in providing key information about a character's personality and background through the provision of initial and limited use of nonstandard features in that person's first spoken utterances. Once a reader has been informed of the message provided by the nonstandard features, it is no longer necessary to repeat such linguistic traits in the character's spoken discourse.

4. Conclusion

Gerrard Winstanley's *The Law of Freedom* is recognised as being one of the most important political tracts of seventeenth-century England. The work, published in the form of a pamphlet, envisages a radically different social and political society based on equality of wealth and opportunity. One of the stylistic strategies employed by Winstanley in furtherance of this vision is spoken discourse. In two of the three sections of the work – the epistle dedicatory and the platform itself – spoken discourse is found in the way of monologic utterance and dialogic exchange. The speakers are numerous, including Oliver Cromwell, the spirit of the king, lawyers, church ministers, the people of England, soldiers, and Gerrard Winstanley himself. This multiplicity of voices in the *Law of Freedom* reflects the polyphonic environment that pamphlets at large created in the 1640s. Many contemporaries were so struck, and frequently bewildered, by the extraordinary explosion of printed opinion and polemic in 1640s pamphlets that far from appreciating what was in essence the emergence of the public sphere they stood aghast at what seemed akin to the biblical state of Babel. For these contemporary critics the variety of voices seemed legion, and far from liberating the positive creative energy of Bakhtin's dialogism, such an onslaught of printed polemic seemed more than partly responsible for the decade's military and political mayhem. For example, in a pamphlet of 1648, the writer considers the country:

(21) Divided as far as Hounds and Hares in antipathizing disaffection: Heads divided in opinions, like those of the Serpent Amphibena, one fighting with another, hearts divided, like fire and water, tongues divided, as still in Babel's confusion, hands divided. (*Mercurius Pacificus* 1648: 2)[23]

---

23 The supposed chaos of thought and opinion brought about by the explosion of pamphlet literature is also represented in 1640s broadsides. In *The World is Ruled and Governed by Opinion* (1641), the woodcut shows blind opinion, in the form of a female figure, bestriding a tree with pamphlets hanging from the tree's branches. See Achinstein (1992: 20-22) for a full explanation of the symbolism in the illustration.

In relation to *The Law of Freedom*, and whether the spoken discourse there reflects a rational, carefully considered use of the spoken word, or, instead, a hectic though passionate expression of multiple voices, one must decide in favour of the former. Winstanley exploits spoken discourse for clear functional purposes to promote a deeply-felt though ultimately unrealised political manifesto. In exploiting the possibilities of spoken discourse to further his radical political agenda Winstanley was not unique. Raymond writes that the Levellers' writings were also "punctuated with speeches and bursts of drama" (2003: 229).[24] These writers were often acutely aware of the variegated background of their audience. They were frequently appealing to distinct audiences: the powers that be, and the dispossessed. In his own political work Winstanley uses oral interaction to inform and entertain the latter but never in such a way as to risk alienating the former.

References

Primary sources

A Perfect Diurnall of Some Passages in Parliament 1645. 76, 6-13 January.
Harrington, James 1656. *Oceana.*
Harrington, James 1659. *Valerius and Publicola.*
Hobbes, Thomas 1651. *Leviathan.*
Marprelate, Martin 1588. *Oh Read Over D. John Bridges, for it is Worthy Worke: Or an Epitome.*
Mercurius Aulicus 23-29 June 1644.
Mercurius Pacificus 1648.
Toland, John (ed.) 1700. *The Oceana of James Harrington.*
The Welch Mercury 21-28 October 1643.

24 Under the leadership of John Lilburne, the Levellers denounced such popular grievances as monopolies, tithes, and the excise, and demanded toleration, freedom of speech, annual Parliaments and the rule of law.

The Wandring Whore 1660. 2.
Winstanley, Gerrard 1649. *A Watch-Word to the City of London, and the Armie.*
Winstanley, Gerrard 1652. *The Law of Freedom in a Platform: Or, True Magistracy Restored.*

Secondary sources

Achinstein, Sharon 1992. The Politics of Babel in the English Revolution. In Holstun, James (ed.) *Pamphlet Wars: Prose in the English Revolution.* London: Frank Cass, 14-44.
Alsop, James 2000. Gerrard Winstanley: What Do We Know of His Life? In Bradstock, Andrew (ed.) *Winstanley and the Diggers 1649-1999.* London: Frank Cass, 19-36.
Biber, Douglas 1988. *Variation across Speech and Writing.* Cambridge: Cambridge University Press.
Biber, Douglas / Johansson, Stig / Leech, Geoffrey / Conrad, Susan / Finegan, Edward (eds) 1999. *Longman Grammar of Spoken and Written English.* Harlow: Pearson.
Brownlees, Nicholas 2005. Spoken Discourse in Early English Newspapers. In Raymond, Joad (ed.) *News Networks in Seventeenth Century Britain and Europe.* London: Routledge, 67-83.
Brownlees, Nicholas 2006. Polemic and Propaganda in Civil War News Discourse. In Brownlees, Nicholas (ed.) *News Discourse in Early Modern Britain.* Bern: Peter Lang, 19-42.
Claridge, Claudia 2005. Questions in Early Modern English Pamphlets. *Journal of Historical Pragmatics* 6/1, 133-168.
Corns, Thomas N. 1992. *Uncloistered Virtue: English Political Literature, 1640-1660.* Oxford: Oxford University Press.
Culpeper, Jonathan / Kytö, Merja 1999. Modifying Pragmatic Force: Hedges in Early Modern English Dialogues. In Jucker, Andreas / Fritz, Gerd / Lebsanft, Franz (eds) *Historical Dialogue Analysis.* Amsterdam: Benjamins, 293-312.
Fox, Adam 2000. *Oral and Literate Culture in England 1500-1700.* Oxford: Oxford University Press.

Fowler, Roger 1991. *Language in the News: Discourse and Ideology in the Press*. London: Routledge.

Habermas, Jurgen 1989 [1962]. *The Structural Transformation of the Public Sphere: An Inquiry into a Category of Bourgeois Society*, transl. Burger, Thomas / Lawrence, Frederick. Cambridge: Polity Press.

Hill, Christopher 1983 [1973]. *Winstanley: 'The Law of Freedom' and Other Writings*. Cambridge: Cambridge University Press.

Jucker, Andreas H. 1997. The Discourse Marker 'well' in the History of English. *English Language and Linguistics* 1/1, 91-110.

Jucker, Andreas H. 2004. Contrastive Analysis across Time: Issues in Historical Dialogue Analysis. In Willems, Dominque / Defrancq, Bart / Colleman, Timothy / Dirk, Noël (eds) *Contrastive Analysis in Language: Identifying Linguistic Units of Comparison*. Basingstoke: Palgrave Macmillan, 97-112.

Kryk-Kastovsky, Barbara 2000. Representations of Orality in Early Modern English Trial Records. *Journal of Historical Pragmatics* 1/2, 201-230.

McElligott, Jason 2004. The Politics of Sexual Libel: Royalist Propaganda in the 1640s. *Huntington Library Quarterly* 67/1, 75-99.

Page, Norman 1988 [1973]. *Speech in the English Novel*. Basingstoke: Macmillan.

Raymond, Joad 1996. *The Invention of the Newspaper: English Newsbooks 1641-1649*. Oxford: Clarendon.

Raymond, Joad 1999. The Newspaper, Public Opinion, and the Public Sphere in the Seventeenth Century. In Raymond, Joad (ed.) *News, Newspapers, and Society in Early Modern Britain*. London: Frank Cass, 109-140.

Raymond, Joad. 2003. *Pamphlets and Pamphleteering in Early Modern Britain*. Cambridge: Cambridge University Press.

Sabine, George H. (ed.) 1965. *The Works of Gerrard Winstanley: with an Appendix of Documents relating to the Digger Movement*. New York: Russell & Russell.

Smith, Nigel 1989. *Perfection Proclaimed: Language and Literature in English Radical Religion, 1640-1660*. Oxford: Oxford University Press.

Taavitsainen, Irma 1999. Dialogues in Late Medieval and Early Modern English Medical Writing. In Jucker, Andreas / Fritz, Gerd / Lebsanft, Franz (eds) *Historical Dialogue Analysis*. Amsterdam: Benjamins, 243-268.

Taavitasainen, Irma 2001. Changing Conventions of Writing: the Dynamics of Genres, Text Types and Text Traditions. *European Journal of English Studies* 5/2, 139-150.

Taavitsainen, Irma / Nevanlinna, Saara 1999. "Pills to Purge Melancholy". Nonstandard Elements in *A Dialogue Against the Feuer Pestilence*. In Taavitsainen, Irma / Melchers, Gunnel / Pahta, Päivi (eds). *Writing in Nonstandard English*. Amsterdam: Benjamins, 151-169.

Zaret, David 2000. *Origins of Democratic Culture: Printing, Petitions, and the Public Sphere in Early-Modern England*. Princeton, NJ: Princeton University Press.

DAWN ARCHER

Tracing the Development of Advocacy in Two Nineteenth-century English Trials

1. Introduction

In previous research I have shown that 1640-1760 was an important period of transition in respect to the (cross-) examination process of the English courtroom (Archer 2005): defence lawyers made their first appearance (in the latter part of the seventeenth century, according to my data) and, as their activity in the courtroom grew, so too did the activity of their 'opponents', the prosecution lawyers. Moreover, as defence and prosecution lawyers became more involved in the examination procedure throughout the eighteenth century, judges seemed to become less involved. Indeed, they were the principal examiners of witnesses and defendants at the beginning of our period but, by the mid-eighteenth century, were adopting more of a presiding role. Defendants also became less vocal, preferring to leave their defence to counsel, when present.

The English courtroom of the mid-eighteenth century was still very different from the modern English courtroom, not least because of a felony counsel restriction which prevented defence counsel from 'addressing the court' directly – by stating what their argument was to be in an opening speech or summarising their argument in a closing statement to the jury prior to them giving their verdict (Langbein 2003: 93). This felony restriction was to remain in place until the passing of the Prisoners' Counsel Act in 1836. According to Cairns (1998: 6), the 1836 Act and the resulting

> shift from restricted to full defence by counsel had a dramatic effect on the development of advocacy and adversarial procedure in the criminal trial. [Indeed,] the relationship of the Prisoners' Counsel Act to contemporary penal reform, its importance to the development of advocacy, and its contribution to

> adversarial procedures justifies a major place for the Act in the history of the modern criminal trial.

This chapter will draw on two nineteenth-century trial transcripts – Angus (1808) and Palmer (1856) – as a means of exploring the extent to which the discursive strategies of the defence lawyers changed, if at all, immediately after the introduction of the Prisoners' Counsel Act (1836). Both defendants were indicted for murder: Angus, for the murder of his children's nanny, Margaret Burns, and Palmer for the murder of his friend, John Parsons Cook.

Given the growing interest in the historical courtroom, American and English,[1] I am also using the opportunity afforded by my investigation to address two issues which I believe should be carefully considered when undertaking (diachronic) research on this particular type of domain-specific data. The first issue relates to the discoursal, socio-historical and legal context we need to take account of to ensure that we adequately account for the data. The second issue relates to the validity of a dataset that is based on (cross-) examination sequences of historical trials, when the aim is to better understand the full range of advocacy practices of the nineteenth-century English courtroom (albeit around the time that the Prisoners' Counsel Act was introduced). I begin with a description of my dataset.

2. Data

The Angus (1808) and Palmer (1856) transcripts consist of ca. 20,000 words and 27,000 words respectively, and are part of a corpus of machine-readable nineteenth century trial texts, which has been supplied to me by Professor Merja Kytö (Uppsala University). I have chosen to utilise these particular texts as:

1 For recent work relating to the language practices of the American and English historical courtrooms, see Archer (2002), Doty / Hiltunen (2003), Hiltunen (1996) and Kryk-Kastovsky (2002). Kryk-Kastovsky has also guest-edited a special issue on historical courtroom discourse for the *Journal of Historical Pragmatics* (7/2, 2006).

- they represent trials that occurred shortly prior to/after the passing of the Prisoners' Counsel Act (1836);
- they contain the examination and cross-examination sequences of a number of different witnesses (i.e. eight each), and thus provide ample evidence of interaction between witnesses and prosecution counsel and witnesses and defence counsel;
- they have been annotated, using the same schemes that were used to annotate the trial section of the Sociopragmatic Corpus (1640-1760).[2] As such, they provide an excellent contrast to the trial texts I used when studying the English courtroom of the late Early Modern English period (see Archer 2005). I provide a description of these annotation schemes in 2.1.

2.1. Annotation schemes added to Angus (1808) and Palmer (1856)

Initially, information regarding the age, status, sex and role of both speaker and hearer was added (when known) to the Angus and Palmer texts at the level of the utterance (Archer / Culpeper 2003).[3] The following utterance, taken from Palmer (1856), shows what the annotation scheme looks like *in situ*:

(1) [*Lord CAMPBELL*] <u stfunc="rep" force="p" speaker="s" spid="s8tpalme003" spsex="m" sprole1="j" spstatus="1" spage="9" addressee="s" adid="s8tpalme005" adsex="m" adrole1="m" adstatus="2" adage="9">-- It seems to me that the examination is conducted with perfect fairness. Nothing has been put in the shape of a leading question about anything which has been hitherto doubtful.</u>

In this particular example, a male judge ("j"), identified as "s8tpalme003", is interacting with a male defence lawyer ("m"), identified as "s8tpalme005". Notice, also, that the judge's status ("1")

---

2 The Sociopragmatic Corpus also contains drama texts. The trial section of the corpus contains 12 texts and totals 121,189 words.

3 The development and implementation of Archer / Culpeper's (2003) Sociopragmatic Annotation Scheme was made possible by a grant awarded to Dr. Culpeper by the British Academy (SG-350252).

is deemed to be higher than the lawyer's status ("2") because of his title (i.e., Lord). Both participants are aged 45 years or over ("9").

A second annotation scheme was then added to the Angus and Palmer texts as a means of capturing the form, force and interactional intent of the speaker(s)'s utterances. The *form* field is relatively self-explanatory, constituting the range of question-types (e.g. yes-no, disjunctive, wh-, negative polar, etc.) used in the English historical courtroom.[4] The *force* field reveals a Searlean (1969, 1975) influence without being a strict lexical classification of so-called illocutionary verbs or Searlean speech acts. Indeed, I assign utterances to one (or more) of seven macro categories (see below), which I view as "reasonably accurate approximations of the prototypical instances of verbal behaviour describable by means of the English verbs used as labels" (Verschueren 1999: 131-132). The illustrative examples are taken from the Sociopragmatic Corpus (1640-1760):

- *counsel* = S wants to convey something to A which will help prevent/result in Y [= an event not in A's best interest], e.g. "My advise to you is, that you would put your self upon your Tryal" (Trial of Slingsby, Hewet and Mordant, 1658);
- *question* = S wants A to supply a missing variable by saying / confirming / clarifying something about X [= an action / event / behaviour / person], e.g. "Shall I withdraw?" (Trial of Charles I, 1649);
- *request* = S wants Z [= an action or event] to happen and hopes to do it or to get A to do (or get others to do) it, e.g. "I now humbly crave a copy of my Charge, time to answer to it, and Counsel assigned me." (Trial of Slingsby, Hewet and Mordant, 1658);
- *require* = S wants (and expects) A to do something, even though A may be reluctant, or to do something him/herself, in spite of A's (probable) reluctance, e.g. "My Lord, I demand this, to hear the Commission read" (Trial of Slingsby, Hewet and Mordant, 1658);

4 See Archer (2005: 130-131) for a detailed description (which includes definitions) of the different question-types.

- *sentence* = S wants to signal a legal decision to A/others (that A is being indicted formally, to be legitimately punished for, or being released from the threat of legal punishment), e.g. "this Court doth adjudg, That the said Charls Stuart, as a Tyrant, Traytor, Murtherer, and a publique Enemy, shall be put to Death, by the severing his Head from his Body" (Trial of Charles I, 1649);
- *express* = S wants A (and others) to know his feelings (about X) and/or cause the action or behaviour to stop, e.g. "The L. Fingshall is a Papist, and Clamorris a pestiferous Papist!" (Trial of Macguire, 1645);
- *inform* = S wants to communicate something (about X) to A, e.g. "We will now call Mr. Buckley." (Trial of Francia, 1716).

These macro-categories are designed to capture 'neighbouring speech acts', that is to say, speech acts that interface with one another in the context of the English historical courtroom (cf. Jucker / Taavitsainen's 2000: 70 notion of 'pragmatic space'). Thus, the *counsel* macro-category subsumes *caution*, *warn*, *threaten*, *compel*, *advise*, and so forth, and the *question* macro-category subsumes *ask* (*about*), *inquire* (*into*), *interrogate*, *query*, and so forth. As I am interested in speech acts as functional units – and "speech acts as functional units do not correspond directly to speech act verbs" (Jucker / Taavitsainen 2000: 70) – I utilise evidence of all speech act phenomena (direct and indirect). However, as the force of some utterances can/may remain indeterminate (because of contextual factors such as status, power and discourse sequencing, for example), my design also allows for the inclusion of multiple and, indeed, indeterminate forces (by using the "m" and "p" values respectively).[5]

The *interactional intent* field relates to the position an utterance occupies in the discourse, and its (interactional/organisational/ structural) purpose. As such, it reveals the influence of Sinclair / Coulthard (1975) and also Stenström (1984). Possible values include:

---

5 Some existing Speech Act taxonomies (e.g. Stiles 1992) lack 'unclassifiable' categories. However, the ability to classify and/or capture ambiguity/ indeterminacy is crucial when studying texts that were produced in periods distant from our own (see also Archer *et al.* forthcoming).

- *initiation* = initiating a new exchange by means of an eliciting device. Prototypical examples: question, request, requirement;
- *response* = providing information that has been directly elicited by another participant, usually by responding verbally. Prototypical examples: answer, acceptance, refusal, denial;
- *response-initiation* = responding to a direct elicitation of another participant by using/following it with an eliciting device. Prototypical examples: an answer immediately followed by a request;
- *report* = stating information which has not been directly elicited by another participant. Prototypical examples: statement, explanation;
- *follow up* = providing follow-up/feedback to a preceding utterance in some way. Prototypical examples: comment, evaluation;
- *follow up-initiation* = providing follow-up/feedback to a preceding utterance by using and/or following it with an eliciting device. Prototypical examples: a comment immediately followed by a question.

The extract below, again taken from the Trial of Palmer (1856), provides an illustration of several of the above values. Mr James, a prosecution lawyer, is questioning a witness, Elizabeth Mills. Notice, however, that the witness uses a checking question (categorised as *response-initiation*) before providing a positive response (categorised as *follow up*)[6] to one of Mr James's questions (categorised as *follow up-initiation*):

(2)	MR. JAMES	You said that you left the room and Cook was sitting up, and Jones, I think you said, supporting him, when I was interrupted; was that so?	[initiation]
	MILLS	I believe he was.	[response]
	MR. JAMES	Where did you remain when you left the room?	[follow up-initiation]
	MILLS	After Mr. Palmer came?	[response-initiation]
	MR. JAMES	Yes.	[response]

6 This is categorised as a *follow up*, as it follows Mr James's positive *response* to the witness's checking question "After Mr. Palmer came?".

MILLS	In the landing just outside the door.	[follow up]
MR. JAMES	Did you wait there?	[follow up -initiation]
MILLS	Yes.	[response]

The *interactional intent* field ensures that (i) I do not confuse turns (i.e. what a speaker says as long as s/he holds the floor), with what the speaker *actually does in a turn*, so that (ii) I can account for those instances *when the speaker does more than one thing in the same turn*, such as responding to given information and immediately eliciting new information (see Stenström 1984: 83). As Section 3 reveals, the scheme is such that I can also determine the level of interaction of each participant group in the two trials.

3. Activity of the main participant groups

In the period 1640-1760, the most active participant groups were the witnesses, followed by the Court/judges and the prosecution and defence lawyers (Archer 2005: 166-174). As Table 1 shows, the same participant groups were also amongst the most active in the trials of Angus (1808) and Palmer (1856):

Trial (year)	*Total utterances*	*Participant group*					
		Defence lawyer	*Prosecution lawyer*	*Witness*	*Court/ judge*	*Defendant*	*Misc.*
Angus (1808)	1,963	246 12.53%	664 33.83%	971 49.47%	76 3.87%	4 0.20%	2 0.10%
Palmer (1856)	2,593	483 18.63%	711 27.42%	1,274 49.13%	123 4.74%	- -	2 0.08%

Table 1. Activity of the main participant groups in Angus (1808) and Palmer (1856).

That said, the Court/judges were not as active as their recent ancestors. Indeed, they were responsible for a mere 3.87% of the utterances in the Trial of Angus (1808) and 4.74% of the utterances in the Trial of Palmer (1856), whilst the judges were responsible for 13.5% of the total number of utterances within the 1640-1760 trial data, and the Court was responsible for 9.3% (see Archer 2005: 166). It is interesting to note, however, that the judges' activity in (cross-) examination sequences declined most steeply during the eighteenth century according to my two datasets, that is to say, from 15.9% in the period 1680-1719 to 2.3% in the period 1720-1760.[7] Significantly, this is the period when we first witness the appearance of lawyers for the defence and a corresponding increase in activity on the part of prosecution counsel.[8] Moreover, the function of the judges' questions also appears to have changed dramatically in the eighteenth century, from one that was primarily information-seeking to one that was primarily clarification-seeking. As will become clear (see Section 5), this trend towards a more overt 'mediator' role (Landsman 1990) seems to continue in the Trials of Angus (1808) and Palmer (1856), in spite of the judges'/Court's slight increase in activity in those trials (when compared with their eighteenth-century counterparts in the trial data from the Sociopragmatic Corpus).

The prosecution and defence lawyers were also more active in the trials of Angus (1808) and Palmer (1856) than they had been in the Sociopragmatic Corpus data for the period 1680-1760. Indeed, prosecution lawyers' utterances accounted for 33.83% of the total in the Trial of Angus (1808) and 27.42% of the total in the Trial of Palmer (1856). Similarly, defence lawyers' utterances accounted for 12.53% of the total in the Trial of Angus (1808) and 18.63% of the total in the Trial of Palmer (1856). In the period 1680-1760, their prosecution and defence counterparts utilised only 14.3% and 9.4% of the utterances respectively (see Archer 2005: 166).

The apparent increase in activity exhibited by the prosecution and defence lawyers in the trials of Angus and Palmer is worthy of further consideration, as it suggests a growing involvement of prosecution and defence lawyers in the courtroom of the nineteenth

7 It is worth noting that the activity of the Court increased during the period covered by the Sociopragmatic Corpus, i.e. from 4.9% in the period 1640-1679 to 20% in the period 1720-1760. Yet, much of this increased activity is attributable to one court case – that of the Court Martial of Ambrose (1745), which (as I argue in Archer 2005) may be atypical of the English historical courtroom in general.

8 Defence lawyers were not present in the 1640-1679 trial texts within the Sociopragmatic Corpus.

century when compared to earlier periods. Yet, this preliminary result also seems to suggest an increase in activity amongst the defence lawyers and a corresponding (albeit small) drop in activity amongst the prosecution lawyers after the introduction of the Prisoners' Counsel Act (1836). I would suggest that we view this particular finding with some caution, however, as:

- the prosecution lawyers were still more active than the defence lawyers in the Trial of Palmer (i.e. the trial that occurred after the introduction of the Act);
- the Angus and Palmer texts do not represent full trials. Rather, as previously explained, they capture eight examination and/or cross-examination sequences each. Consequently, a drop in activity amongst the prosecution lawyers might be the result of editorial choices made when constructing the nineteenth-century trial corpus (i.e. the decision to choose one examination sequence over another).

For this reason, I will be concentrating on the sort of activity the various participant groups were involved in (rather than their level of activity) from this point onwards. Moreover, as the main aim of this chapter is to explore the extent to which the strategies of the defence lawyers changed (if at all) after the introduction of the Prisoners' Counsel Act, I will be focussing, mainly, on the defence lawyers' interaction with other participant groups in the Trial of Palmer (1856).

4. A breakdown of the utterances of the defence lawyers

The majority of the defence lawyers' utterances – i.e. 217 (+22) out of 246 in Angus and 421 (+39) out of 483 in Palmer – had a questioning force.[9] Most of these questions (i.e. 97.97% in the case of the Angus

---

9 The bracketed figures relate to utterances with multiple forces. Readers should note that 22 of the defence lawyers' utterances had a multiple force in the

defence lawyers and 98.14% in the case of the Palmer defence lawyers) were addressed to witnesses, with only a few utterances being addressed to other participant groups (i.e. the Court in the case of the Angus defence lawyers, and the judge and prosecution lawyers in the case of the Palmer defence lawyers) – see Table 2.

Participant group	*Angus defence lawyers (in %)*	*Palmer defence lawyers (in %)*
Witness	97.97	98.14
Prosecution lawyer	0	1.24
Court/judge	2.03	0.21
Problematic	0	0.41

Table 2. Participant groups questioned by defence counsel in Angus and Palmer.

In Archer (2005), I suggest that the defence lawyers' strategies were becoming more controlling in the late Early Modern English period (this was especially true of the eighteenth-century trial texts I studied). By *controlling*, I mean that defence lawyers were increasingly utilising question types such as (negative) yes/no interrogatives and declaratives, which narrowed the choice of answers available to the respondent and/or allowed the examiner rather than the respondent to phrase the evidence (see Archer 2005: 151, 213-221). Indeed, a third (i.e. 33.5%) of the interrogatives utilised by the 1680-1760 defence lawyers had a yes/no structure, whilst negative yes/no interrogatives and declarative questions accounted for 4.6% and 4.9% of their total questions respectively (see Table 3). Defence lawyers in the trials of Angus (1808) and Palmer (1856) also utilised yes/no interrogatives and declarative questions (see Table 3 for percentages). Notice, however, that the Palmer defence lawyers seemed to prefer the yes/no interrogative, whilst the Angus defence lawyers seemed to prefer the declarative question, which would suggest that the Palmer lawyers were less controlling of witnesses than the Angus lawyers.[10]

Angus text, and 41 of the defence lawyers' utterances had a multiple force in the Palmer text.

10 Note that Table 3 makes use of an amended version of Woodbury's (1984: 205) 'continuum of control'. The continuum is designed to show the different levels of control exhibited by the various question types. However, as the amount of control exhibited by particular question-types – and thus the level

Least controlling/ Low conductivity	*1680-1760 SPC trials*		*Angus (1808)*		*Palmer (1856)*	
wh-	92	(28.6%)	20	(8.1%)	118	(24.9%)
disjunctive	13	(4.0%)	3	(1.2%)	5	(1.0%)
yes/no	112	(33.5%)	57	(23.1%)	229	(48.3%)
Negative yes/no	15	(4.6%)	12	(4.9%)	0	(-)
Declarative	16	(4.9%)	123	(49.8%)	82	(17.3%)
Tagged declarative	0	(-)	11	(4.5%)	4	(0.8%)
Most controlling / high conductivity	Total q-types = 325		247		474	

Table 3. Question-types utilised by the defence lawyers, and their level of conducivity.

It would be worth undertaking more extensive research of the nineteenth-century English courtroom to determine (i) the extent to which the higher use of conducive/controlling strategies in the Angus trial is indicative of trials prior to 1836 in general, and (ii) whether a direct link can be made between defence lawyers' inability to address the jury and the use of highly conducive questioning strategies. I would advise against assessing conducivity/control on interrogative structure alone, however, for question types that are non-conducive or relatively conducive can become more conducive in the courtroom. As the following extract from the Trial of Palmer (1856) reveals, this is especially true when non-conducive or relatively conducive question types are utilised as part of a questioning sequence which contains other question types that are highly conducive:

(3) [Context: Gibson has informed the prosecution counsel that he treated Cook – Palmer's apparent victim – giving him an anodyne draught and paregoric]

Cross-examined by Mr. Serjeant Shee.
SHEE Did you form an opinion what was the matter with him?

---

of restriction they impose upon the respondent's options – seems to correlate with the strength of the questioner's expectation of an answer, I have supplemented Woodbury's continuum within a *conducivity* scale.

GIBSON	I went to work and treated it as if he had taken poison.
SHEE	You took him at his word?
GIBSON	Yes.
SHEE	You did not observe anything in what he vomited which should lead you to think so?
GIBSON	Nothing at all in what he vomited.
SHEE	Did you observe any symptoms from which you could infer that he had been taking any poison?
GIBSON	He had pain.
SHEE	That is not very uncommon, is it?
GIBSON	No.

Notice, for example, that the opening question here is a yes/no question. However, it is part of a questioning sequence that utilises declaratives and tagged declaratives. Indeed, the yes/no interrogative sets up a (recurrent) I-R-F/I sequence that allows the lawyer to intimate the witness had not been as diligent as he might have been in his treatment of Cook.

5. The Court's use of questions (Angus 1808)

The above extract suggests that lawyers were able to utilise questions in role-specific ways, in this case, to undermine (i) the testimony given by a witness, and (ii) the actions of a witness under questioning and/or at the time of the alleged crime. My data suggests that lawyers were not the only participant group to utilise questions in such role-specific ways. Indeed, the Court in the Trial of Angus (1808) often adopted the role of clarifier and/or mediator (see 4 and 5 respectively):

(4)	COURT	She did not observe her Master, the first time she went into the room?	
	TOPPING	No, my lord, she only just opened the door.	[inform]
	COURT	She observed her Master the second time, but as you put the question, it seems as if she had seen him both times.	
	TOPPING	No, my Lord, I am inquiring about a different thing – this was when she first came down in the morning.	[inform]
	COURT	Her account before, was, that when she had opened the door and saw something, as she says, cower'd down, she was frightened, and ran	

		into the kitchen, and asked the Cook where Miss Burns was – then she went back with the Cook, and then it was, that she discovered her Master lying on the chair.
(5)	HOLROYD	When he mentioned her having taken a black puke, what was the expression?
	SCARLET	My Lord, I must really object to this – I have not cross-examined about that, and the question is, therefore, irregular. [express / inform]
	COURT	You have made so much confusion, between the black puke, the emetic, and the tartar, that I declare it is almost unintelligible.
	SCARLET	Why, my Lord, if there is any doubt upon that subject, I should certainly wish to have the privilege of enquiring into it, in my cross-examination. [request]

Interestingly, the Court in the Angus trial adopted the role of clarifier and/or mediator with the defence counsel slightly more than it did with the prosecution counsel (i.e. on four occasions as opposed to one, according to my data). Analysis of more trials would enable us to determine whether judges generally interacted more with defence lawyers than they did with prosecution lawyers in the nineteenth century. That study might also seek to determine the extent to which the irritated tone discernable in some of the clarifications addressed by the Court to the defence lawyers in the Angus trial (see the Court's response to Scarlet's objection in (5), for example) is also characteristic of judges' interaction with defence lawyers at this time and, if so, what type of effect this may have had on the jury. I would argue that the answer to this last question (regarding the effect on the jury) could prove to be extremely informative, especially given that the recognised role of the judge at that time was to "prepare the minds of the Jury for correctly forming their own conclusion" rather than to "urge them to find a particular verdict" (Cairns 1998: 111).

6. The lawyers' tussles in the *Trial of Palmer* (1856)

Defence lawyers in the Trial of Palmer (1856) interacted very little with the Court/judges during (cross-) examination sequences (but see

Section 7.2.2). Indeed, they only addressed one utterance to the judges – "no my Lord" – which functioned as a response. The Palmer defence lawyers also interacted very little with the Palmer prosecution lawyers during their (cross-) examination sequences, i.e. 6 utterances (out of a possible 483). However, their interaction with the prosecution lawyers is very interesting, as the majority of the utterances (i.e. four of the six) functioned as complaints about / objections to the prosecution counsel's use of "suggestion" and/or requests that the practice stop (see 6 and 7 respectively):

(6) MR. JAMES (to witness) After the beating had gone on with his hands and arms, as you have described, did you observe any stiffness about his arms or hands?
WITNESS He asked me to rub one hand.
SHEE I must request my learned friend to ask the witness what she observed and not to suggest. [request/express]
MR. JAMES I must ask you to object to any question which I put.
SHEE I object to my learned friend suggesting any symptom; let the witness state the symptoms which she observed: that suggestion was very important [express; objection]

(7) MR. JAMES When you went into the room, in what state was Mr. Cook; was Mr. Jones still with him?
WITNESS Yes, he was.
MR. JAMES Was he sitting up in the bed?
WITNESS Yes, much the same as he was when I left.
SHEE I must ask my learned friend again not to suggest; he is suggesting: "Was he sitting up in the bed?" The moment before he asked, "Did he say anything about his dress?" There is a suggestion every instant. [express; objection]

As will become clear (Section 7.2.2), this tussling between the lawyers was also a feature of the defence counsel's opening speech in the Trial of Palmer (1856).

7. The advent of the Prisoners' Counsel Act (1836)

The merits of the felony counsel restriction were hotly debated before, during and immediately after the enactment of the Prisoners' Counsel Act in 1836. Defenders of the felony counsel restriction would often point out that the Act was not necessary, as prosecution counsel generally exercised a degree of self-imposed restraint because defence counsel were not permitted to address the jury (May 2003: 87). In contrast, the proponents of full defence by counsel insisted that a system that allowed a speech to one counsel and not to the other was always going to be unfair, even if those giving a speech exercised restraint (Cairns 1998: 99). What was not in dispute by either side was the belief that, once full defence by counsel was introduced, the contest between prosecution and defence counsel would become much more adversarial.

By the time of the Palmer trial in 1856, the debate was much less fierce but things had changed, not least because the English criminal trial was now characterised by "a philosophy which sought truth in the contention of counsel" (Cairns 1998: 106). Indeed, one contemporary – Fitzjames Stephen – described the post-1836 English criminal trial as "a public inquiry, having for its object the discovery of truth, but thrown for the purpose of obtaining that end into the form of a litigation between the prosecutor and the prisoner" (1863: 166). A leading writer on English criminal law, Stephen was present at the Palmer trial. Given his credentials, it is worth noting that he described the trial itself as "exhibit[ing] in its very best and strongest light the good side of English Criminal procedure" (Stephen 1890: 269, quoted in Cairns 1998: 164). He, like many of his peers, was particularly impressed with the advocacy of the chief prosecution counsel, Sir Alexander Cockburn, for he was said to have presented the Jury

> with a picture of Palmer's guilt in his opening which he maintained and reinforced throughout the trial. It guided the examination of witnesses, cross-examination, and is even apparent in his re-examinations, for so fine is the advocacy in this case. (Cairns 1998: 162)

The next section explores some of the linguistic features of Cockburn's opening speech.[11]

7.1. The linguistic features of the Attorney-General's opening speech

After jury selection and some other preliminaries, Cockburn made his opening address to the jury. As guilt or innocence could not be based on opinion, Cockburn was careful to emphasise the *duty* that he and the jurors faced during the opening moments of his speech, and particularly emphasised the need for the jurors to base their findings on the *evidence* only (italics are mine):

(8) Gentlemen of the jury, the *duty* you are called upon to discharge is the most solemn which a man can possibly have to perform – it is to sit in judgment and to decide an issue on which depends the life of a fellow human being who stands charged with the highest crime for which a man can be arraigned before a worldly tribunal [...]. *Your duty – your bounden duty – is to try this case according to the evidence which shall be brought before you, and according to that alone.* You must discard from your minds anything that you may have read or heard, or any opinion that you may have formed. If the *evidence* shall satisfy you of the prisoner's guilt, you will discharge your *duty* to society, to your consciences, and to the oaths which you have taken, by fearlessly pronouncing your verdict accordingly, but if the *evidence* fail to produce a reasonable conviction of guilt in your minds, God forbid that the scale of justice should be inclined against the prisoner by anything of prejudice or preconceived opinion [...]. My *duty*, gentlemen, will be a simple one. It will be to lay before you the facts on which the prosecution is based.

The Attorney General's advice to the jurors that they should base their findings on the evidence presented to them is not surprising, given the context. However, the reader should note that the prosecution's case against Palmer was actually based on a certain type of evidence – circumstantial evidence – for no trace of the poison that the prosecution argued was used to murder Cook was found in his remains. The word *evidence* is also interesting from a linguistic point of view, for the Attorney General uses the word much less than Shee

11 As the Palmer text I am utilising only captures (cross-) examination sequences, I have taken the opening speeches of the respective counsel from Watson's (1952) revised edition of *The Trial of William Palmer.*

does in his opening speech for the defence (i.e. 14 times as opposed to 118). As raw figures can be misleading when the texts we are comparing are not the same size (as in this case), I have undertaken a keywords analysis of the opening speeches of the prosecution and the defence in the Trial of Palmer (1856), using WordSmith Tools (Scott 1999).[12] I report my findings for the Attorney General's opening speech in Section 7.1.1, and my findings for Shee's speech in Section 7.2.1.

7.1.1. A keyword analysis of the Attorney General's speech

A keyword analysis comparing the Attorney General's speech with that of Serjeant Shee has revealed that *evidence* was statistically infrequent in the former's speech, as was *tetanus*, *letter*, *opinion*, *know*, *friend*, *gentlemen*, *if* and *not*. The words that were statistically frequent included *strychnine*, *clock*, *Palmer*, *Cheshire*, *sickness*, *prisoner*, *motives*, *forged*, *crime*, etc. A closer examination of some of the positive keywords (i.e. the words that were statistically frequent in the Attorney General's speech when compared to the opening speech of Shee) reveals specific semantic fields relating to:

- (the symptoms/outcome of) sickness, and the treatment of (e.g. *sickness*, *disease*, *fever*, *sick*, *vomiting*, *antinomy*, *pills*, *doctor*, *bilious*, *medicine*, *consciousness*, *strychnine*, *deceased*, *fatal*);
- (periods of) time (e.g. *clock*, *period*, *September*, *January*, *hours*, *Thursday*, *year*, *evening*, *days*, *day*);
- crime (e.g. *forged*, *commit*, *crime*);
- people (e.g. *Bamford*, *Cheshire*, *Padwick*, *Bates*, *Hawkings*, *Newton*, *women*, *chambermaid*, *Boycott*, *Gardner*);
- places (e.g. *Tattersall*, *Shrewsbury*).

The above keywords and the semantic fields to which they belong suggest that the Attorney General was careful to emphasise the victim

12 This analysis allows the identification of words in a text that are statistically (in)frequent when compared against a normative corpus. In order to ensure a certain degree of result validity, I set the *probability* at the typical value for the social sciences (i.e. ≤0.05, which equates to a 5% chance or less of the results being a fluke).

(Cook) suffered before he died in his opening speech. He also seems to have been specific about the times and places in which the alleged crime took place, and also specific about (i) the people who were involved in the crime itself (i.e. *Palmer*), (ii) the people whose activities were said by the Attorney General to trigger the alleged crime (i.e. *Cheshire*), and (iii) the people who treated/cared for Cook (i.e. *women*, *chambermaid*, *Newton* and *Bamford*).

This type of narrative – described as a criminal occurrence narrative by Gergen (1999: 69) – is very common in opening speeches made by prosecution counsel today. It is also aptly named, as such narratives tend to be interpretative as well as informative. The interpretive element of the Attorney General's opening speech was at its most obvious when he attempted to account for the accused's intentions, by establishing a motive for Palmer's apparent actions. Keywords that help to construct *motive* in the Attorney General's opening speech are *liabilities*, *money*, *bills*, *forged*, *betted*, *debt* and *turf*. Indeed, cumulatively, they allowed the Attorney General to propose that Palmer's gambling debts were such that he initially forged his mother's name to secure a new gambling income. However, when that source of funding dried up and his debts began to mount again, he resorted to murder. As we shall see (Sections 7.2-7.2.1), Serjeant Shee painted a very different picture in his narrative.

7.2. The linguistic features of Shee's opening speech for the defence

The first six days of the Palmer trial were taken up by the prosecution counsel, who initially introduced their narrative in their opening speech and then reinforced/expounded upon that narrative during the examination of 66 *friendly* witnesses (but see Section 7.2.3). Shee therefore had to wait until day seven to give his opening speech, on behalf of the defence. That speech is unusual for a number of reasons. First, Shee chose to declare his personal belief in his client's innocence in very explicit terms. He did so by initially highlighting that he, "like everybody else [...] knew generally and loosely [...] the history of these transactions at Rugeley" and by so doing, highlighted the infamy (and perhaps the prejudicial views against Palmer) that the case had generated prior to coming to trial. Shee then stressed how, having allowed himself "no opinion upon the guilt or the innocence of the prisoner at the bar" prior to studying the papers that had come into his hands, he had since been able to reach "the right judgement in this case", which was that Palmer was innocent. Indeed, he was so convinced that he proclaimed "I believe that there never was a truer word pronounced than the words which he [Palmer] pronounced when he said 'Not guilty' to the charge". Such behaviour may not seem scandalous outside of the courtroom context (indeed, we might think it admirable). However, as the Lord Chief Justice in the Palmer trial made clear in his summing up, within the courtroom context it was seen to "press [...] his opinion" as opposed to pressing "his argument upon the jury" (see Watson 1952: 297-298). Indeed, by going beyond the role of the prisoner's representative in this way, Shee managed to "challenge accepted norms of forensic morality" and also "the division of responsibility between the judge, witnesses and counsel, which was the basis of adversial procedure" (Cairns 1998: 155).

Second, and relatedly, the Attorney General and the Lord Chief Justice consistently intervened in Shee's opening speech. Before we explore those interventions in any detail, I will briefly report on the keywords within Shee's opening speech.

7.2.1. A keyword analysis of the opening speech of the defence

The bulk of Shee's statistically frequent words – i.e. *evidence*, *know*, *opinion*, *innocence*, *improbable*, *probable*, *supposed*, *inconsistent*, *suggest*, *credible*, *suggestion*, *belief*, *conceive*, *actually*, *justify*, *supposition*, *if*, *clear*, *would* and *could* – were utilised to emphasise the importance of evaluating the evidence of the case correctly, that is to say, in a way that exonerated Palmer. Shee did not seek to offer the jury a complete counter-narrative, however, but instead sought to undermine those aspects of the opposition's narrative that implicated Palmer directly.[13] For example, he emphasised the friendship between Palmer and Cook, as a means of ridiculing the idea that Palmer had

13 Contextual factors may account for Shee's choice of strategies, as he was having to respond to six days worth of prosecution testimony. He was also retained by Palmer only shortly before the case came to trial (see Cairns 1998).

poisoned Cook for his money. Indeed, Shee stated of Palmer that he was:

(9) the dearest and best friend of Cook [...] who liked him and loved him well enough [that] when he [...] was ill [he was prepared] to sleep with him in the same room that he might be ready to attend to him in case he wanted assistance during the night.

Additional keywords within the friendship domain utilised by Shee in his opening speech are *friends* and *affection*.

As the statistical relevance of *correspondence* highlights, Shee also made much of the fact that, in his estimation, the papers relating to the case actually pointed to "the innocence of the prisoner". Indeed, he suggested that, if one was to look at the "long correspondence [that] has been put in […] as a matter of business and scrutinise it in every corner […][then] this correspondence saves the prisoner, if there is common sense in man". Part of this correspondence also related to *letters* (another keyword in Shee's opening speech) that were used by the prosecution to prove that Palmer was in debt and, by so doing, establish his motive for murdering Cook. But Shee utilised those same letters to highlight that Palmer was helped by his friend and helped his friend in return. He also quoted from another letter (dated 4th December) that admitted doctors had failed to "find strychnine, prussic acid, or any trace of opium" in Cook's remains.

A third feature of Shee's opening speech worthy of mention is the use he made of the first person (related keywords include *my*, *I* and *we*). For example, a concordance search of *my* reveals *learned friend* and *duty* as strong collocations. The former was usually employed in reference to the Attorney General. Shee utilised *duty* in much the same way as the Attorney General did in his opening speech (which helps to explain why *duty* was not a statistically (in)frequent word in either Shee's speech or the Attorney General's speech). *Ability* also collocated with *my* in Shee's opening speech. In respect to *my ability*, Shee was signalling the effort he would exert in defending Palmer, especially given the fact that he had been detained by him at such a late stage:

(10) Gentlemen, one great misfortune has befallen the accused – a most able man who had been selected by him as his counsel many weeks ago has been,

> unfortunately, by illness prevented from discharging that duty to him. I have endeavoured, to the utmost of my ability, to supply his place; I cannot deny that I am awed – that I am moved – by the task I have undertaken; but the circumstances to which I have already adverted, the national effort, so to speak, through the Government of the country, to ensure a fair trial is a great cause of encouragement, and I am not dismayed.

In respect to Shee's use of *I* and *we*, several usages were not self-referential as we might expect but, rather, captured the *voices* of different players in the unfolding drama. For example, Shee explained how:

(11) We had Mr Deane called yesterday, who is the attorney to the Prince of Wales insurance office; [...][that office was] annoyed at being called upon to pay so large a sum [of insurance that] they determined to do all they could to resist it. They accordingly sent down Inspector Field to Stafford and his man Simpson to make inquiries, which he could not do without talking and insinuating suspicions and raising a cloud of doubt and conjecture about Palmer, and this had been going on for some considerable time. Now, observe the evidence of Deane, and you will see if it is not so. He says, "The name of my firm is Chubb, Deane & Chubb. I had been to Rugeley some time previously to the inquest. I know Field, the detective officer; we were solicitors to the Prince of Wales insurance office; it was in our employment that Field went to Rugeley; he was at Rugeley only a part of one day; he was at Stafford for three or four days altogether; he did not see the prisoner Palmer; this visit had been preceded by that of another officer named Simpson."

Shee then went on to report both Deane's account of Simpson's visit with Palmer and Deane's assumption that "just before the death of Cook Palmer knew himself to be an object of suspicion" as a means of being able to argue that "by insinuations and inquiries [...][they] destroy[ed] his character". The use of direct speech seems to have been strategic on Shee's part, as he consistently reconstructed the words of other non-friendly witnesses, in particular, so that their words became self-condemning and/or might have prompted the jury to re-imagine particular scenes from a perspective more favourable to Palmer. In the case of Shee's re-use of Deane's words (see above), it was to provoke the jury to recognise Palmer as a victim (of "insinuations and inquiries") and not a murderer (see Section 7.2.2).

When Shee's use of the first person was self-referential, pronouns often collocated with *believe* and *think*, thereby introducing a more subtle expression of personal opinion which, unlike his earlier declaration (in his belief in Palmer's innocence), was not as heavily criticised by the Attorney General and the Lord Chief Justice. It seemed, then, that lawyers could offer their personal opinion, as long as it was not done explicitly. Indeed, to do so was deemed to have merged functions (i.e. testimonial and defensive) that were meant to be kept separate. For you could be accused of

> assum[ing] the role of judge or jury in reaching a decision on the innocence of [your] client, and then at the trial testifying in support of a not guilty verdict.
> (Cairns 1998: 158)

Such explicit declarations of personal belief also tempted prosecution counsel to respond in kind. Interestingly, the Attorney General utilised his own implicit expression of personal opinion to signal that he would not succumb to such a temptation:

(12) I can only say I think it would have been better if my learned friend had abstained from so strange a declaration. What would he think of me if, imitating his example, I at this moment stated to you, upon my 'honour', as he did, what is my internal conviction from a conscientious consideration of this case. The best reproof which I can administer to my learned friend is to abstain from imitating so dangerous an example.

7.2.2. Shee's opening speech: monologue or dialogue?

The Attorney General did not choose to intervene in Shee's opening speech to criticise him for his use of personal opinion, but he did intervene to object to evidence that Shee had introduced without sufficient 'proof'. He also rebuked the defence counsel for introducing evidence which he "would not let [the Attorney-General] ask him [Palmer] any questions about", and acknowledged a mistake made by the Attorney General and corrected by Shee: "You are quite right as to the £500".

The Lord Chief Justice also intervened in Shee's opening speech. The purpose of three of these interventions was to correct Shee's explanation of certain facts: the first related to Shee's assessment of a witness's statement in relation to tetanus, the second to Cook's state of health on a certain night and the third to Shee's assessment that the prosecution had made too much of Palmer "pushing" a doctor whilst involved in Cook's post-mortem examination. The purpose of the fourth intervention was to verbally acknowledge that Shee could "read [...] only that part of the deposition which describes the symptoms".

Given opening speeches are regarded as monologic today, it would be worth undertaking further research of opening speeches of the nineteenth century to determine whether they, too, were monologic in the main or, like the Palmer trial, exhibited characteristics of dialogue. That research might consider both the direct intervention of others and also the reconstruction/inclusion of other voices, via the primary speaker's use of direct reporting.

7.2.3. Tracing the opening speeches in the examination sequences

Using the annotation schemes applied to the Angus (1808) and Palmer (1856) texts, I identified specific stretches of interaction between the Attorney General and Shee and their respective witnesses, and then performed a third and fourth keyword analysis using WordSmith Tools. Length constraints prevent a detailed analysis of all the keywords here. Suffice it to say, I found that some of the positive keywords in the Attorney General's examinations (*found*, *health*, *period* and *symptoms*) were also statistically frequent in his opening speech. Similarly, some keywords in his examinations (*health, dose, convulsions, convulsive, muscles*, *tissues* and *horse*) were not key in his opening speech but did belong to the medical and motive/gambling semantic fields which we have identified as being important to the latter (see Section 7.1.1).

Surprisingly, none of the other semantic fields we identified were statistically frequent in the Attorney General's examination sequences. However, Shee did utilise positive keywords relating to the people and time domains when questioning witnesses (for example, *Cook's*, *coroner*, *Dolly's*, *mother*, *Stephens*, *Dutton*, *Barnes*, *Lavinia*, *Talbot* and *yesterday*, *before*, *now*, *times*). Like the Attorney General, Shee also had an identifiable medical domain (*coroner*, *throat*, *complain*, *vomited*). This particular domain is indicative of the stress that the defence counsel placed upon the lack of strychnine in Cook's remains. However, what Shee did not seem to do was ask questions of witnesses that might lead to a natural explanation for Cook's sickness and ultimate death. Instead, the bulk of the positive keywords (e.g. *swear*, *asked*, *talk*, *spoke*, *stated* and *complain*) suggest that he focussed on the witnesses' verbal activity (both around the time of the alleged event and also during the trial itself). In the following interaction, for example, Shee asked a series of questions which were designed to ascertain what certain people discussed when they met together:

(13) On this occasion was all the talk about Mr. Cook's death?
- I cannot remember; it might be mentioned.
Do you mean to say there was no word said about anything else?
- Yes, I dare say there was.
Will you swear there was a single subject of conversation mentioned?
- I do not pretend to keep in my head what the conversation was.
Will you undertake to say there was a single subject of conversation mooted between you and Lavinia Barnes and those gentlemen except the subject of Cook's death?
- Will I undertake to say it, do you say?
Yes.
- There were many more things talked about.
What?
- That I do not wish to mention.
You must mention what was the subject of conversation?
- I cannot remember; it will be impossible for me to remember.
Do you mean to undertake to swear these gentlemen did talk to you on other subjects besides the death of Mr. Cook?
- They were not talking with me alone, but among themselves; I was not paying attention to what they were talking about.

As the extract makes clear, the witness (Elizabeth Mills) was not very forthcoming. Consequently, Shee was forced to be much more direct about the evidence he was really seeking:

(14) Did they talk about the trial that was coming on?
- That I do not know.
Did they talk about the evidence that you were to give?
- No; not that I heard.
Did they ask you what you could prove?
- No.
Did they read their depositions over to you, taken before the coroner?
- No, they did not.
Did they read anything else to you?
- I saw nothing produced.

> Was anything read to you?
> - No.
> Was anything read to you from a newspaper?
> - No.

If Mills had answered positively, the prosecution case would have been severely damaged. As it was, Shee was able to insinuate that his client really had been the victim of "insinuations and inquiries", as he had claimed in his opening speech (see Section 7.2.1).

Significantly, Shee underused the other dialogic strategy he had used in his opening speech (the first person singular *I*). Yet, in spite of being statistically infrequent (when compared to the Attorney General's examination sequences), some first person singulars still referred to the voices of others. For example, Shee asked the same witness, Elizabeth Mills, "Did you not tell him [Stephens] after he had been calling two or three times, 'Why, Mr. Stephens, you have been here often enough; I have told you all I know'?".

The above keywords study is incomplete, of course, as the Angus (1808) and Palmer (1856) texts only capture the interaction between the Attorney General, Shee and the witnesses in the Palmer examination sequences within the nineteenth-century corpus. It is possible, therefore, that Shee and the Attorney General picked up on other themes from their respective opening speeches in additional examination sequences not captured by my version of Palmer. It is also worth remembering that both Shee and the Attorney General were one of a team of lawyers. And one or all of those lawyers may have focussed on additional themes initially highlighted by the Attorney General and/or Shee in their interactions. The incomplete nature of the Angus and Palmer texts also raises a second important issue, that of validity, which I will address in the final section of this study.

8. Developing a richer understanding of the courtroom

I have sought to highlight a number of prominent linguistic features in the two trial texts, Angus (1808) and Palmer (1856), and, by so doing, put forward a number of preliminary hypotheses that require further research if they are to be (in)validated. For example, I have highlighted an apparent move from more conducive question-types to less conducive question-types, and suggested that we might engage in further research to establish whether this trend is characteristic of nineteenth-century English trial texts in general as well as the Angus and Palmer texts in particular. That research might also seek to determine if a less conducive questioning strategy is related at all to the introduction of opening speeches for the defence and, importantly, whether it should be intrepreted as being less controlling of witnesses. This is especially important, given contemporaries seemed to have argued that courtroom battles became *more* adversarial after the enactment of the Prisoners' Counsels Act (Stephen 1890).

As we have seen, that Act widened the role of the defence counsel, for both sides could now present their version of events to the jury directly, and then confirm, develop or contest these narratives during their examination sequences with witnesses. But, twenty years on, Shee's behaviour in the Palmer trial suggests that he, at least, was either uncertain as to how far he could go in encouraging jurors to "discern the truth" or, alternatively, deliberately introduced personal opinion in order to test the limits of forensic argument at that time (Cairns 1998: 161). Further research is required to determine the extent to which counsel within other nineteenth-century trials also drew on personal beliefs in their opening speeches and closing arguments and/or attempted to put their words into the mouths of the witnesses during their (cross-) examination sequences.

If we are to undertake such research, we must begin (where possible) to include complete transcripts of individual trials within corpora rather than representative extracts (as is usually the case at present), so that we have access to lawyers' (cross-) examination sequences, opening speeches and closing arguments (as well as the verbal activities of the other participant groups). I would also suggest that we engage in socio-historical and legal research, so that we better appreciate the working practices and the etiquette of the English historical courtroom. Such studies will reward us, in turn, with a much improved understanding of the rich complexities of this particular language domain from a diachronic perspective.

References

Archer, Dawn 2002. "Can innocent people be guilty?" A Sociopragmatic Analysis of Examination Transcripts from the Salem Witchcraft Trials. *Journal of Historical Pragmatics* 3/1, 1-30.

Archer, Dawn 2005. *Questions and Answers in the English Courtroom (1640-1760): A Sociopragmatic Analysis.* Amsterdam: Benjamins.

Archer, Dawn / Culpeper, Jonathan 2003. Sociopragmatic Annotation: New Directions and Possibilities in Historical Corpus Linguistics. In Wilson, Andrew / Rayson, Paul / McEnery, Tony (eds) *Corpus Linguistics by the Lune: Studies in Honour of Geoffrey Leech.* Frankfurt: Peter Lang, 37-58.

Archer, Dawn / Culpeper, Jonathan / Davies, Matthew Forthcoming. Pragmatic Annotation. In Lüdeling, Anke / Kytö, Merja / McEnery, Tony (eds) *Corpus Linguistics: An International Handbook.* Berlin: Mouton.

Cairns, David 1998. *Advocacy and the Making of the Adversarial Criminal Trial 1800-1865.* Oxford: Clarendon.

Doty, Kathleen / Hiltunen, Risto 2003. "I will tell, I will tell". Confessional Patterns in the Salem Witchcraft Trials, 1692. *Journal of Historical Pragmatics* 3/2, 299-335.

Gergen, Kenneth J. 1999. *An Invitation to Social Construction.* Beverly Hills: Sage.

Hiltunen, Risto 1996. "Tell me, be you a witch?" Questions in the Salem Witchcraft Trials of 1692. *International Journal for the Semiotics of Law* 9, 17-37.

Jucker, Andreas H. / Taavitsainen, Irma 2000. Diachronic Speech Act Analysis: Insults from Flyting to Flaming. *Journal of Historical Pragmatics* 2/1, 67-95.

Kryk-Kastovsky, Barbara 2002. *Synchronic and Diachronic Investigations in Pragmatics*. Poznań: Motivex.

Kryk-Kastovsky, Barbara (ed.) 2006. Historical Court Trial Discourse. Special issue of *Journal of Historical Pragmatics* 7/2.

Langbein, John H. 2003. *The Origins of the Adversary Criminal Trial.* Oxford: Oxford University Press.

Landsman, Stephen 1990. The Rise of the Contentious Spirit: Adversary Procedure in Eighteenth-century England. *Cornell Law Review* 50, 498-609.

May, Allyson N. 2003. *The Bar and the Old Bailey 1750-1850.* Chapel Hill, NC: University of North Carolina Press.

Scott, Mike 1999. *WordSmith Tools.* Oxford: Oxford University Press.

Searle, John R. 1969. *Speech Acts.* Cambridge: Cambridge University Press.

Searle, John R. 1975. Indirect Speech Acts. In Cole, Peter / Morgan, Jerry (eds) *Syntax and Semantics. Volume 3: Speech Acts.* New York: Academic Press, 59-82.

Sinclair, John M. / Coulthard, Richard M. 1975. *Towards an Analysis of Discourse: The English Used by Teachers and Pupils.* London: Oxford University Press.

Stenström, Anna-Brita 1984. *Questions and Responses in English Conversation.* Malmö: Liber.

Stephen, Fitzjames 1863/^{2}1890. *A General View of the Criminal Law.* London.

Stiles, William B. 1992. *Describing Talk: A Taxonomy of Verbal Response Modes.* Beverly Hills: Sage.

Verschueren, Jef 1999. *Understanding Pragmatics.* London: Arnold.

Watson, Eric R. 1952. *Trial of William Palmer.* Revised Edition. London: Hodge.

PAOLA TORNAGHI

William Dugdale and MS Harley 1129: An Unpublished Seventeenth-century Legal Glossary

1. Introduction

This chapter is part of a research project on OE lexicography in the sixteenth and seventeenth centuries, when complex cultural, political and religious events were arousing interest in the retrieval of the linguistic and cultural heritage of the Anglo-Saxons and the Middle Ages. This interest was to result in the birth and growth of English philology, associated with the identification of a Germanic matrix in English language and culture. It was mainly during the times of Henry VIII and Elizabeth I that ties between the ruling dynasty and the medieval past were unearthed, and a new interest developed in the Anglo-Saxons – their history, laws and language. As the language was quasi-'dead', glossaries, dictionaries and grammars were necessary for scholars to be able to read and interpret the manuscripts – see, among others, Lawrence Nowell's *Vocabularium Saxonicum*, John Joscelyn's *Dictionarium Saxonico-Latinum* (MSS Cotton Vitellius A XV and XVI), Sir Symonds D'Ewes's *Dictionarium citeriorum Saeculorum Anglo-Saxonicum Latinum* (MSS Harley 8 and 9) and William Dugdale's *Dictionarium Saxonico-Anglicum* (MSS Harley 8 and 9).[1]

While consulting William Hamper's volume (1827) on William Dugdale's life, diary and correspondence at the British Library, I came across a footnote which mentioned MS Harley 1129. Intrigued, I followed it up at the Dept. of Manuscripts and found that, although the

1 The first of these lexicons was published in 1952 by Marckwardt; the others, still unpublished, have already been analysed and transcribed under my supervision within the research project mentioned before.

manuscript was not autographed, the handwriting was undoubtedly William Dugdale's typically small, neat hand.[2]

I will not dwell on the historico-cultural background which brought about the revival of studies on the Anglo-Saxon world in the sixteenth and seventeenth centuries, nor will I give an overview of the Anglo-Saxon lexicographic production, which was always marked by close collaboration among the so-called antiquarians and by their constant interdependency and generosity in sharing any information they were able to retrieve.[3] What I intend to stress is that, although many seventeenth-century antiquarians were involved in the enterprise of recovering the Middle Ages (with John Selden and Sir Henry Spelman in a leading position), William Dugdale (1605-1686) stood out as the great retriever of English antiquities and he pursued his aims with tenacity. Anthony Wood (1692, quoted in Parry 1995: 219) concludes Dugdale's autobiography[4] emphasising the most telling aspects of the antiquarian's personality and work. His "true and accurate" studies saved a cultural inheritance which otherwise would have been lost to posterity:

> Had this indefatigable person sequestred himself from wordly troubles, and totally addicted himself to his studies, and had minded the public, more than his private concerns, the world might have justly enjoyed more of his lucubrations, and those more true and accurate, than such that are already published, especially those in his latter days: Yet however what he hath done is prodigious, considering the great troubles that he had endured for his loyalty, and the cumbrances of this world that he had run through, and therefore his memory ought to be venerated and had in everlasting remembrance, for those things which he hath already published, which otherwise might have perished and been eternally buried in oblivion.

---

2 The (still partial) results of the research I present here are also due to the precious collaboration of Francesca Barantani, who transcribed the contents of the MS under my supervision.

3 On this subject see also, most recently, Parry (1995) and Tornaghi (1999).

4 Dugdale himself wrote his autobiography towards the end of his life. His friend Anthony Wood had access to this document and printed much of it in his notice of Dugdale in *Fasti Oxonienses* of 1692, adding material to extend it to Dugdale's death. The autobiography was reprinted together with Dugdale's diary and correspondence in Hamper (1827).

From this kind of epitaph the fact that much of Dugdale's work was never published comes to light. Since his studies were based on constant research, selection and consultation of ancient documents, they were always *in fieri*, always perfectible. It is also likely that this antiquarian regarded his research and his writings as the result of an activity of secondary importance, something like a 'hobby'.

What impresses about Dugdale is indeed his prodigious output – both published and still manuscript – and his multifarious activity as a 'literate knight' and a lexicographer, especially in relation to his *Dictionarium Saxonico-Anglicum*, compiled in 1644 (Oxford, Bodleian Library MS Dudgale 29), to his contribution to the publication of William Somner's *Dictionarium Saxonico-Latino-Anglicum* (1659) and of the second volume of Spelman's *Glossarium Archaiologicum* (1664), and eventually to the glossary of legal terms found in MS Harley 1129.

2. MS Harley 1129

In the Catalogue of the Harleian Manuscripts (Nares 1808-1812) the following description of MS 1129 is given:

> A Book in folio given by the Revd. Doctor George Hickes, late Dean of Worcester; wherein is written, A kind of Glossary, wherein many obscure Words observed by Sir William Dugdale in his Reading are noted down, as also many Authorities explaining the same, all by his own hand. He had divided the Book into fit Portions, according to the Order of the Alphabet; and left Room for inserting more Words as occasion should require.

A footnote in William Dugdale's biography, published by Hamper (1827: 37), reads:

> The Harleian MS 1129 contains Alphabetical Collections by Dugdale for a Glossary of Ancient Law Terms, and other words of base Latinity, on 100 Folio pages in double columns.

These short descriptions give important information about the manuscript: it is clearly attributed to Sir William Dugdale and the contents are specified. It is in-folio, written both on the recto and the verso; only the recto pages are numbered with the number placed in the top right-hand corner. There is a total of 342 headwords on 105 folios.

Each page is ruled for two columns on each side, separated by two vertical lines down the middle and flanked by a top-to-bottom line both on the right and the left. As a heading, each column has a sequence of two capital letters corresponding to the initial letters of the headwords contained in each column (for example AB, AB, AC, AC, AD, AD etc.). The number of entries varies from column to column and some columns have no entries at all: the manuscript must have been structured so that headwords could be added. Entries are often ordered randomly under the column-headings; alphabetical order is not strictly followed because Dugdale tended to arrange and group words which presented similar forms and meanings. In addition, words starting with <u> or <v> are not ordered alphabetically, since the two graphemes are used indifferently to express either /u/ or /v/: *ucha* follows *vavassor*, *ulna* follows *virgata terrae*.

Each entry consists of the lemma, its definition and the quotation from the relative source: examples are given below, with emphasis added in boldface, to highlight the lemma. The source is described in detail as to title, author, chapter, page, sometimes the publishing year and even the place where the source is to be found (3 and 8). The source is usually presented in a shortened form. The definition is generally written in Latin (3), English (1 and 4), Latin and English (5 and 6), but we find also definitions written in Latin and OE (7), and in a few cases also Latin and French (2), according to the language of the source itself:

(1) **Aelmesfeoh**. or Almesmoney, was the Peter pence due yearely at the first of August by institution as some will of King Ina, as others of King Æthelulph, and they were called also Romefeoh, Romescot, and Heorþpeninȝ. Seld[en] Hist[orie] of Tithes p: 217[5] (f.6^{v})

---

5 Underlined letters represent the expansion of shortened words in the MS.

(2) **Baro**. [...] [In margine sic annotatur. Barones[6] pris pour gentilhommes à qui le seignour de Talmont avoit accorde des fiefs en l'estendue de sa principauté] Les preuues de l'Hist[oire] des Comtes de Poictou (Œ) p. 322 (f. 13r)

(3) **Cambiarius** mansit tamen ibi in medio agri (Œ) quidam ceruisiae brasiator vel cambiarius ubi rustici homines et incompositi ad bibendum vel ad cheolandum vel etiam herkandum[7] propter agri pascui largam et latam planiciem convenire solebant. Les preuues de l'Hist: des comtes de Guines. Imp. Pariss. 1631. p. 142. (f. 15r)

(4) **Cotarii, or Cotucami, or Coterelli**. Cotagers. Cokes Inst: part 1. fol:5.b[8] (f.18r)

(5) **Ferlingus sive Ferlingum**, or Quarantena terrae. Is a furlong of land, & is as much as to say a furrow long which in antient tyme was the eight part of a myle. Cokes Inst: part 1. fol: 5.b (f. 24r)

(6) **Managium**. a house or dwelling place a mansion —— Concessi capitale Managium meum cum pertinentiis, (Carta Nich: de Balitiam) (f. 41r)

(7) **Minister regis** Sax: cyninȝes ðeȝn. Seld: tit: of honour. Pag. 611. et 687. (f. 43r)

(8) **Swillinga**. id est hida. Seld: tit: of Honour p: 636 MS. Oxon: in biblioteca: corporis Christi T. annot: fol: 48.b. (f. 81r)

Sometimes Dugdale uses curly brackets to group the meanings of a single lemma or, vice versa, a list of lemmas with the same meaning:

(9) **Hoo** } a Hill. Cokes Inst[itutes] part: 1 fol: 4.b et 5.b(f. 32r)
Howe

Curly brackets are also used to group citations from different sources, in which the same word is recorded, e.g. *socne* (see Appendix).

A different layout appears when Dugdale quotes an explanatory passage from a source describing a dispute between two people over a

6 Underlined in the MS.

7 *Cambiarius, cheolandum, herkandum* are underlined in the MS.

8 The source reads: [...] *cotucami cotarij* are all in effects bores or husbandmen, or cotagers. (Latin, *cotarius,* coterus) The cottagers were the smallest of the four major groups among the peasantry. For more detail, see Maitland (1897), Lennard (1959), Darby (1977), and Dyer (1985).

heritage: the different versions of the two parties are placed in two distinct columns and collected under the same entry:

(10) **Astrarius** haeres inde dicitur ille cui antecessor in vita sua (per cartam) hereditatem restituit. Cokes Instit[utes] part 1 fol: 8.b
Galfridus tempore regis Johannis =

Willelmus filius antenatus =	Robertus de Albe postnatus =
Galfridus de Albe petens qui dicit quod praedictus Robertus non fuit nec potuit esse astrarius praedicti Galfridi, nec unquam remansit de praedicta terra tamquam, astrarius sed dicit quod idem Galfridus per longum tempus ante mortem suam dederat ei duas carucatas terrae in Drayton ad quam terram idem Robertus duxerat uxorem, et fuit separatus à familia Galfridi patris sui.	Alexander de Albe defendens dicit quod Robertus pater suus, et filius praedicti Galfridi remansit in praedictis terris et messuagio et redditibus, ut filius eius et astrarius sed perjuratus fuit manens apud Drayton, et penitus separatos ab astro patris sui.

Placita de banco termino Mich. Circiter annum 36 vel 37. H.3. Rot: 21. Leic¯. in Ontheby. (f. 10^{r})

Some entries refer to other words recorded in the same manuscript by means of cross-references:

(11) **Hacton**. vide Acton (f. 29^{r})
Biannum. vide Bannum. vide Tallita (f. 13^{r})
Hercandum. vide Cambiarius (f. 30^{v})

Others refer to words recorded in different dictionaries, such as Cotgrave's or Spelman's, which are used as sources. This proves that Dugdale was well aware of the lexicons compiled by his "antiquarian colleagues":

(12) **Bacon**. (both French & English. vide Cotgr[ave] Dict[ionary]) (f. 11^{v})
Custrill. Id est Coustillier. Vide Cotgraves Dict. (f. 18^{v})
Quarentena terrae. vide Ferlingus. vide Spelman gloss: verbo Leuca. (f. 57^{v})

Dugdale records many graphemic variants (e.g. *Ætheling, Adeling, Edeling, Frescenga vel Friscinga, Herciare, vel Herseare, Lea or Ley, Leuuad, Leuga, Leuued & Leuue, Warectum or Wareccum, or Varectum*) and uses many abbreviations both for Latin and English, which are recurrent in manuscripts. As for handwriting, Dugdale does not differentiate Latin from English; only some Anglo-Saxon words are written in the typical Anglo-Saxon insular cursive hand. Sometimes, when explaining a lemma, he underlines some words, either Latin or English, because they also occur or should have been recorded as specific entries. With reference to (3), where *cheolandum* and *herkandum* are underlined in the manuscript, we see that *cheolare* and *hercandum* are respectively recorded as follows:

(13) **Cheolare**. (fortaβe idem est quod manducare). vide Cambiarius
Hercandum. vide Cambiarius (f. 30ᵛ)

As for the date when this glossary was compiled, no precise indication is provided.[9] However, it is reasonable to presume that the glossary was compiled in the period following Dugdale's main lexicographic work, the *Dictionarium Saxonico-Anglicum*, in 1644.

The outbreak of the Civil War in 1642 must have disrupted Dugdale's studies. His affiliation to the College of Arms placed him in the Royalist camp. His four-year stay with the Court, at Oxford, until June 1646 when it surrendered, gave him the opportunity to search the archive collections of the Bodleian Library and the Oxford colleges. In 1646 Dugdale moved back to London, where he continued to collaborate with Roger Dodsworth at the compilation of *Monasticon Anglicanum* and to expand his research aimed at the realisation of his other project, a history of Warwickshire, his father's adopted county. The work, for which he is now best known, was published in 1656 under the title *The Antiquities of Warwickshire.* It was not the first county history to have been written but in its day it was certainly the best and most detailed. In 1666 Dugdale published another historical work entitled *Origines Juridiciales, or Historical Memorials of the English Laws, Courts of Justice, Forms of Tryall, Punishment in*

9 On the last page there is a date, February 1878, when the Manuscript is likely to have been purchased and catalogued by the British Library.

Cases Criminal, Laws Writers, Law Books, Grants and Settlement of Estates, Degree of Serjeant, Innes of Court and Chancery, of which he wrote in his biography: "having in his many yeares labours in search of Records for those workes already published, taken Notes he had taken of the Lord Chancelours, Lord Treasurers, Masters of the Rolls [...] he compiled that historicall Worke, intituled Origines Juridiciales" (Hamper 1827: 28). In 1676 this work was followed by *The Baronage of England*, a landmark in the history of English genealogical scholarship (Hamper 1827: 5-42; Hetherington 1980: 88-92; Parry 1995: 217-247; www.britannica.com/eb/article-9031379).

During the perusal of all the material at his disposal for the compilation of these works, Dugdale must have had the opportunity to take note of legal terms and those peculiar terms connected with legal matters, which are recorded in the glossary under scrutiny.

3. Comparing sources

> As then by our Ancestors the actions of former times are thus left to us we alsoe may leaue to posteritye those of the present. (Hamper 1827: 150)

Dugdale thought that as a historian he should record events as faithfully and impartially as possible. As an expert antiquarian, Dugdale applied this principle to his lexicographic research. That is why each entry of his glossary is accompanied by detailed information about the source which contains the definition explaining the lemma itself with the greatest degree of precision. Dugdale's accuracy allows "posteritye" to check the correctness of his indications and compare his notation of the *interpretamentum* and of the source.

Where possible, the text of the cited source has been compared in order to check the lexicographer's absolute adherence to it; see for instance the examples below:

(14) **Adeling**. id est domicellus. R. Hou: f: 347ª n° 40
___ sed nos indiscrete de pluribus dominis, quia Barones vocamos domicellos: Angli vero nullum nisi natos regum, quod si expressius volumus dicere in Saxonica lingua imago dicitur Adela Angliće, Nobilis Latine, quod

simul coniunctum sonat nobilis imago Adeling, unde etiam occidentales Saxones Cestrenses habent in prouerbio summi despectus Edeling, id est omni honestate deiectus vel retrocedens imago. Ib: n° 50. (f. 6^{r})

The source Dugdale cites is Rogerus de Hoveden's (†1201) work, *Chronica Magistri Rogeri de Hoveden*, in which we read:

> *Quomodo Sancta Margareta, mater regis David Scotiae, cum patre et matre, sorore, et fratre Edgardo Edelingo, de quo Sanctus Edwardus rex heredem regni sui facere voluit, in Angliam venit.* [...][10] Et quia cogitabat eum heredem facere, nominavit eum **Adeling, quod nos dicimus *domicellum*. Sed nos indiscrete de pluribus dominis, quia barones vocamus domicellos; Angli vero nullum nisi natos regum. Quod si expressius volumus dicere, in Saxonica, *ling* imago dicitur, *Adela* Anglice, *nobilis* Latine. Quod simul conjunctum sonat, *nobilis imago*, Edeling. Unde etiam occidentales Saxones Cestrenses [Execestre] habent in proverbio summi despectus Edeling, id est, ab omni honestate dejectus, vel retrocedens imago.**
>
> (Stubbs 1964: 236)[11]

(15) **Sulh:ælmessan**. Plough=Almes, w<u>hi</u>ch was a penny to be paid of every plough land. Seldens Hist: of Tithes p. 222.vide W: Malm:de gest:Regum Angl: lib : 2. f. 42.a n° 10 (f. 80^{r})

In John Selden's *Historie of Tithes*, page 222, we find:

> Those *Aratrales Elemosynae* were called **sulh ælmessan**, that is *Plough-almes*; **which was a peny to be paid of euery plough-land**.

(16) **Arbergaria**. Et ita volo (₵) ut nullus meorum (₵) cogat sibi praebere arbergariam, aut hospicium, aut quaerat ab eis <u>quod</u> talliatam vocant. Preuues de l'Hist. de Comtes de Poictou. (₵) p.367

The source reads as follows:

> **Et ita volo** eos esse quietos, scilicet monachos, et omnes res eorum, hominumque eorum vbilibet positorum, **vt nullus meorum** non filius non filia, non uxor non aliquis propinquus, non Dapifer, non Praepositus, non Mariscalcus, non seruiens, aut in aliquo ministerio positus, aut monachos

---

10 As above, boldface highlights the lemma and the text quoted by Dugdale. Quite a long passage from the cited source has been omitted here because it does not occur in Dugdale's quotation.

11 See also <www. gallica.bnf.fr/scripts/ConsultationTout.exe?O=N050282>.

iamdicti Monasterii, aut homines eorum in quocumque loco eorum habitent, **cogat sibi praebere arbergariam, aut hospitium, aut quaerat ab eis quod talliatam vocant** (Besly 1647: 367).

As we can see, Dugdale selects the passage he considers useful and suitable to define the entries appropriately. He employs graphic symbols like <₢> signalling omissions (16) and often quotes more than one source (15).

In addition, he often records the author's name and the title of the work in abbreviated form. These abbreviations, however, are sometimes difficult to interpret, especially as Dugdale is not always consistent, and often employs different abbreviations for both the authors and the works he quotes. For example, John Leslie, bishop of Ross (1526-1596), is quoted four times in four different ways:

(17) I: Lesl scotus episcopus Ross: De origine moribus et rebus gestis Scotorum (*s.v.* **Acton**)
J: Lesl: Scot: de orig. moribus et rebus gestis Scotorum (*s.v.* **Brache**)
Lesl:episcopus Ross: de orig moribus et rebus gestis Scotorum (*s.v.* **Pailes**)
J.Lesl: Scot: de orig: morib: et rebus gestis Scot: (*s.v.* **Thainus Regis**)

In Leslie's case, the different abbreviations have not raised particular problems in identifying the source. On the contrary, the case of the Scottish philosopher, Hector Boece (1465-1536), is more complicated because he is mentioned twice in the whole manuscript with two different abbreviations without any reference to a specific work:

(18) Hect: B: (*s.v.* **Acton**)
Hect: Boet: (*s.v.* **Werdas**)

Here it is interesting to observe that there is a discrepancy in the quotation recorded by Dugdale and the source text we have consulted: the lemma *Acton* refers to both John Leslie and to Hector Boece:

(19) **Acton.**
___ Lorica hamis ferreis conserta muniebantur, hanc tunicae coriaceae non minus firmae, quam eleganti (nostri Acton dicunt) superinduerunt. Johannis Lesl: scotus. episcopus Ross: De origine moribus et rebus gestis Scotorum. pag. 56
___ Scoti venandi, aucupandique arte sese excercebant, pascebantque greges. In his eorum opes arcu missilibus, levi armatura, loricis, alii ferreis, alii

> pelliceis, vulgo **Hactones** (sic) dicunt, vsi non aliter – quam aperto Marte spem posuerunt libertatis simul et regni tutandi. Hect: B: f. 5.a. (f. 5^{r})

In *Scotorum Historia a prima gentis origine cum aliarum et rerum et gentium illustratione non vulgari, libri XIX* we read:

> Scoti venandi aucupandique arte sese exercebant, pascebantque greges, in his eorum opes, arcu missilibus, leui armatura, loricis, alii ferreis, alii pelliceis, vulgo **Nactones** dicunt, vsi non aliter quam aperto Marte spem posuerunt libertatis simul & regni tutandi. (Boece 1527: f. 5a)

The headwords recorded by Dugdale are *Acton*[12] and *Hacton*. His definition in the glossary records *Hactones*, not *Nactones* as appears in the source. Dugdale may have misunderstood the graphemes <h> e <n>, but the difference is more likely to be ascribed to a printing mistake in the text by Boece (1527). Such a hypothesis would be supported by the fact that in Du Cange's (1883-1887) medieval Latin dictionary what is recorded is the headword *Acton / Aketon* and not the entry *Nacton*.

The most recurrent source quoted by Dugdale is *Coke's Institutes* (1628), which appears 110 times in the glossary; it is a four-volume collection of English laws, which are considered the basis of the modern English legal system, and are part of the legislative heritage known as Common Law. Another source which is frequently quoted is *Les Preuues de l'Histoire des Comtes de Poictou et ducs de Guyenne* (44 occurrences); it is included in a work entitled *Histoire des Comtes de Poictou et ducs de Guyenne*, written by Jean Besly and published in Paris in 1647. Dugdale refers to the *Preuues*, a collection of mostly Latin documents that were employed by Besly to draw up his work and were published as relevant documentation to the book itself. They were documents of different kinds, taken from old chronicles or collected in churches and monasteries. They must have contained legal charters or notarial acts in which Dugdale found analogies with terms belonging to the English medieval law. Dugdale is likely to have consulted or even purchased this volume during his

12 In medieval Europe *acton* was a racket or a jerkin originally of quilted cotton worn under a coat of mail. It also denoted a leather racket padded with mail (Collins English Dictionary ^{6}2003, *s.v.*).

stay in Paris in June 1648,[13] a year after the publication of Besly's work.

Among English legal works are John Selden's *The Historie of Tithes*, published in London in 1618, quoted 13 times, and *Titles of Honour*, published in London in 1614 and 1631, quoted 9 times. The former deals with the history of tithes in England, the latter lists and comments on the meaning and value of ranks and titles.

Other sources deal with English and Scottish history.[14] Many registers, charters and documents from abbeys and monasteries are also cited and most of them must have been consulted by Dugdale when he collaborated with Dodsworth on the compilation of the *Monasticon Anglicanum*, an account of English monastic houses, consisting, to a large extent, of charters of foundation and other original documents.[15] Finally, there are ten entries which are not followed by any source:

(20) **Firmaculum**. a claspe or buckle[16] (f. 25^{r})
Palefridus. a Horse for travaile, of French word Palfray (f. 48^{v})
Pelvis cum lavatorio. a basin and Quire (f. 50^{r})
Runcinus. a Horse for burthen or labour. It cometh of the Italian word Roncino id est a colt. (f. 64^{v})
Ruche, Gallica lingua significat A Bee hive (f. 64^{v})
Serif. vicecomes. Scotice (f. 71^{r})
Solarium a foller, vpper room or chamber. Dedi (Œ) vnam shoppam cum Solario superedificato. (f. 71^{r})
Supertunica a Surcoate, or vpper garment. (f. 80^{r})

---

13 Dugdale went to France in 1648 owing to the collapse of the Royalist cause.

14 For English history we have Rogerus de Hoveden's *Chronica Magistri Rogeri de Hovedene*, William of Malmesbury's *De Gestis Regum Anglorum* and Henry of Huntington's *Historia Anglorum* (all published in London in 1596). For Scottish history we have John Leslie's *De Origine Moribus et Rebus Gestis Scotorum* (Rome 1578), Hector Boece's *Scotorum Historia a prima gentis origine* (Paris 1527), and James Buchanan's *Rerum Scoticarum Historia* (Edinburgh 1582).

15 The first volume of *Monasticon* appeared in 1655, the year after Dodsworth's death and just seventeen years after the authors began their joint work. The second volume, which was delayed until the sale of the first should produce funds to defray some of the expense, came out in 1661; in 1673 Dugdale published a third volume containing *Additamenta* and documents relating to the foundation of cathedral and collegiate churches.

16 There are some cancelled words; "a kinde of dreßing" is legible.

Viches, Furrura de viches. The furr of a vermin called a Tichow. (f. 95^{v})
Uuastina, idem quod vastum. Anglice was grounde. (f. 101^{r})

This shows that the glossary was being constantly updated with headwords which were not specifically legal, but must have been included in legal documents and used in legal contexts. Though never finished, the glossary possibly followed Dugdale right through his life.

4. A captivating legal historical account

Rather than a dictionary to be consulted, MS Harley 1129 is a work to be read as if it were an interesting historical account in three languages (indeed, four, if we consider OE a separate language), written without pedantry, and aiming to recount a lost world.

Dugdale's lexicons are the result of a patient, meticulous collection of historical, legal, religious and scientific material taken from a variety of manuscripts and other contemporary works. A distinctive trait of Dugdale as a lexicographer is his painstaking accuracy, which depended upon his commitment as an antiquarian, keen on history and particularly aware of the importance of sources, and as a herald, whose accuracy was essential for the reliable description of monuments, coats of arms and suits of armour or the reconstruction of family trees. As a herald, Dugdale also had the opportunity to use many ancient Anglo-Saxon charters, wills and records of property transactions. Indeed, Dugdale's correspondence, published by William Hamper in 1827, is an important source of information on his activity as a lexicographer and the contacts he kept with his contemporaries. For instance, letter 31, addressed to Sir Symonds D'Ewes and dated February 27th, 1639 reads:

> I received the Peterborough bookes and Mr Allen's booke the last weeke. I will not fayle to doe my best endeavour for you against Easter Terme to gett Sr Wm Peshales Deedes and that other of Basett which you desire. I will deliver Mr Elsings Booke to Mr Johnson sometyme next weeke. I am in despaire to obteyne the bookes of Sir Thomas Cotton which you desire. I am glad to heare of your inclination to print the Saxon Dictionarye, and especially

> the English therewith, inasmuch as it hath soe much consonancye to the Saxon. (Hamper 1827: 195)

Dugdale was in touch with many noteworthy antiquarians of the time, such as Dodsworth, D'Ewes, Somner, Spelman, Cotton, Junius, Blount and others. What emerges from the letter above is that these scholars exchanged not only books, manuscripts, and documents, but also notes, comments and suggestions. They encouraged one another to pursue their objectives. The publication of a work was often the result of the collaboration and the research of several scholars. All this was crucial in the compilation of lexicographical works. In his letters, Dugdale suggests how to interpret a number of obscure English terms and points to possible sources of information. However, Dugdale never refers to the compilation of a glossary or of a dictionary of his own. His secrecy is perhaps due to the fact that he did not consider his lexicographic research as relevant as, for example, D'Ewes's and Somner's – and he did not see his results as complete works worthy of publication.

A hint at his *Dictionarium Saxonico-Anglicum* can be found in two letters written by Roger Dodsworth in Spring 1650, in which the publication of Dugdale's work is mentioned (Hamper 1827: 226, 229). We do not know what Dugdale replied, but Dodsworth's words prove that Dugdale compiled the dictionary and that D'Ewes and Somner could consult it. Possible hints at the compilation of MS Harley 1129 can also be found in some letters Dugdale wrote to Thomas Blount (1618-1679), who, as a Catholic, could not practise as a lawyer, and devoted himself to the study of law and the English language and to the compilation of 'hard word dictionaries'. In 1656 he had published his first dictionary, *Glossographia*[17] *or, a dictionary interpreting the hard words of whatsoever language, now used in our refined English tongue*, and 1670 saw the publication of a dictionary of legal terms, *Nomolexicon: a law dictionary interpreting such difficult and obscure*

17 This work not only aimed to present a complete listing of English words, but also to define and explain unusual terms that might be encountered in literature and in the professions. It went through several editions and remains amusing and instructive reading. With ca. 11,000 entries, it was the largest English dictionary when it was published.

words and terms as are found either in our common or statute, ancient or modern lawes.

In a letter to Sir John Cotton, Sir Robert Cotton's son, dated October 15th, 1675, Dugdale recommends Blount as a "person well verst in Antiquities, and Author of that usefull *Law Dictionary* (for interpretation of old difficult words) [...] for which respect he well deserves all encouragement in this his commendable Studies" (Hamper 1827: 401). As usual, Dugdale offers Blount his great experience as an antiquarian and gives him helpful suggestions. In February 1678 he writes to him:

> and that you may see I do often thinke of you I have here sent you a very pertinent exposition of the word *Rageman*, which neither Sir Henry Spelman nor yourself have met with. (Hamper 1827: 420)

Dugdale knew Spelman's work[18] very well and must have read Blount's lexicon very carefully. Thus he realised that neither lexicographer had recorded the term *Rageman*. Dugdale readily puts the results of his personal research at Blount's disposal, providing his own interpretation and the source he had consulted at the end of the same letter:

> Anno 1274 – The same year was one *Ragimund*, a Legate, directed hither; who called before him al the Beneficed persons within the kingdome, caused them upon their oath to give up the worth and value of their Benefices, according to which they were taxed. The table, commonly called *Ragimund's Roll*,[19] served for the present collection, and was a Rule in aftertimes, for the prizes taken of those who came to sue for Benefices in the Court of Rome –

---

18 Spelman (1562-1641) is well known for his detailed collections of medieval records, in particular of church councils. His histories include *Concilia Ecclesiastica Orbis Britannici* and *Glossarium Archaiologicum* (completed by William Dugdale). His *Reliquiae Spelmannianae* was edited by Edmund Gibson in 1698.

19 Ragman's-Roll, n. A roll or register of the value of benefices in Scotland, made by Ragimund, a legate of the pope, according to which the clergy were afterwards taxed by the court of Rome. It was later applied to the four great rolls of parchment recording the acts of fealty and homage done by the Scots nobility to Edward I. in 1296; these four rolls consisted of thirty-five pieces sewn together. The originals perished, but a record of them is preserved in the Rolls House, Chancery Lane.

History of the Church of Scotland. By Archbishop Spotswood, Imp. London 1655, pag. 46. (Hamper 1827: 422)

In Dugdale's glossary the entry *Lot* reads as follows:

(21) **Lot**. Praesentatum est in Regemannis per xii de alto Pecco quo Ragimundus le Wyne fecit quandam purpresturam in solo domini Regis in Talington a Prasteliue faciendo mineram plumbi, unde Rex solebat percipere le Lot[20] minerarium· id est tertium decimum [?][21](Œ): Rotuli: Rageman. et de Quo waranto de itinere de Derb: 9°. E. 1.
—— Et de minera lucrata in huiusmodi opere in feodo domini Regis, dominus Rex habebit pro dominio suo tertium decimum discum qui dicitur le Loth. Escaet: de anno 16°. E. 1. n° 34 Derb: (f.39^{v})

Here both *Rageman* and the name *Ragimund* occur together with the mention of the *Rotuli Ragemannis*, proving that Dugdale knew the word *Rageman* very well, as he had already recorded it in his glossary.

Both MS Dugdale 29, containing the *Dictionarium Saxonico-Anglicum*, and MS Harley 1129, show that Dugdale, despite his intense activity as a herald and a politician, must have devoted much time and in-depth research to the compilation of these lexicons. His habitual association and friendship with many of the most important antiquarians of his time – and many of them were also lexicographers – coupled with his innate passion for history and the Anglo-Saxon world must have further stimulated his studies in lexicography.

This kind of research must have involved him so much that in the already mentioned letter to Sir Symonds D'Ewes, dated February 27th, 1639, he writes:

I thank you much for your great respect to me, and in particular for that peece of Saxon which I hope to receive from you. Sir Henry Spelman hath given me a Saxon Psalter, and I purpose, God willinge, to bestow some tyme in that study hopinge to obteine some small portion of knowledge therin.
(Hamper 1827: 195)

"To obteine some small portion of knowledge therin" seems to be Dugdale's main purpose, and his commitment as a lexicographer

20 Both *Lot* and *Loth* are underlined in the MS.
21 Illegible word in the MS.

ought to be considered in this light, which would explain the reason why his lexicographic works remained unpublished and his biography does not openly refer to them.

Compared to Dugdale's *Dictionarium Saxonico-Anglicum*, MS Harley 1129 has different contents, form and structure. However, both manuscripts present some common characteristics which can be regarded as typical of the way Dugdale used to work and substantiate his great ability as a scholar of antiquities.

The method with which Dugdale records and lists entries is crucial – his extreme accuracy in the choice of entries, in their definitions and, in the case of this manuscript, accuracy in the retrieval of sources, has earned him a permanent place amongst the country's foremost antiquarians both as a historian and an expert in legal matters and as a lexicographer.

Re-reading MS Harley 1129 as a whole we realize that Dugdale was not only thorough when carrying out his research and honest when recording all the information retrieved, but he was also brilliant when narrating stories about specific entries, and clever when choosing what to tell. We may wonder why he included an entry such as *Venerabilis Beda* in a glossary of legal terms, but we immediately understand why when we read the amusing anecdote in Latin:

(22) **Venerabilis Beda**. Quare sic dicebatur? Ragimundus Remington de gestis regum Angliae volens ostendere quam ob rem Beda dicebatur venerabilis. Fuit Romae, inquit, porta ferrea in qua hae literae summatim erant scripta. P.P.P. S.S.S. R.R.R. F.F.F. quas nemo legere poterat vel intelligere. Postea Bedae Romam venienti et easdem literas inspicienti, dixit quidam Romanus Quid aspicis Anglice bos. Cui Beda. Confusionem vestram aspicio. Igitur Beda quod in eisdem literis latebat in hoc modo declarabat. Pater patriae perditus est. Sapientia secum sublata est. Ruet regnum Romae. Ferro flamma fame Quod autem audiens et omnis populus Romanus, dignum iudicaverunt honore, et Venerabilem[22] censuerunt appellare. I Caius de antiq: Cantab: Acad: p. 104. (f. 95^{r})

As a matter of fact, Bede is still widely known as 'venerable'. Through this story, Dudgale intends to offer an explanation for this epithet.

---

22 Underlined in the MS.

We may also wonder why he recorded a word such as *trebuchet* by defining it as "a cuckingstoole" and referring to Coke's *Institutes* without quoting the specific context of the source:

(23) **Trebuchet**. a cuckingstoole. vide Cokes Instit: part 3 pag: 219

We are, thus, invited to read the source, which clearly says that it was an instrument of torture and explains how it worked:

> Trebuchet or castigatory [...] signifieth a Cucking stool, and Trebuchet properly is a pitfall or downfall, and in law signifieth a stool, that falleth down into a pit of water, for the punishment of the party in it.

These two words are still in use in contemporary English, and dictionaries record that *trebuchet* or *trebucket* is a large medieval siege engine for hurling missiles consisting of a sling on a pivoted arm set in motion by the fall of a weight. It dates back to thirteenth century. The word sounded obsolete in the sixteenth century and was later recorded as historical and antiquarian by legal antiquaries[23] (OED, *s.v.*; Collins English Dictionary ^{6}2003, *s.v.*). As for *cucking-stool*, it is also known as "Stool of Repentance" and is of very ancient date (OED, *s.v.*; Collins English Dictionary ^{6}2003, *s.v.*)

The two terms are not synonyms, but they must have been related owing to the similar shape of the two objects. We may assume that Dugdale was perfectly aware that the reader knew about this variety of catapults and this specific kind of instrument of torture.

Many entries in the Glossary refer to Latin words, at times also French, for names of plants and herbs, such as *Arundinetum, Lupulicetum, Salicetum, Feugere, Iaun* etc. The question is why these entries are recorded in a glossary of legal terms. One of the main sources used by Dugdale, *Coke's Institutes,* presents an entire section at the beginning, which is dedicated to the recording of such terms; although pertaining to botany, they needed to be defined and explained whenever they occurred in legal documents (for instance, in contracts, disputes over borders or land expropriations, deeds of transfers etc.).

---

23 Cf. Cotgrave, 1611, *Trebuchet*, a pitfall for birds; a pit, with a trap doore, for wild beasts; also, a paire of gold weights; also, an old-fashioned Engine of wood, from which great, and battering stones were most violently throwne.

Dugdale collected and recorded all the 'legal' terms and those connected with legal contexts which may have sounded peculiar, difficult and/or obsolete, selecting them from sources he presumably employed daily.

When reading through the glossary, we realise that Latin is more widely used than English. There are also French quotations. As is well known, Latin was the official language in the legal field. Anglo-Saxon law derived from Roman law and, although many enlightened kings (such as Æthelberht, Offa and Alfred) wanted their law codes to be written in Anglo-Saxon, during the long period of Anglo-Norman and French rule, laws were written in Latin and French. Besides, in Dudgale's days Latin was the language of culture and governance and most scholars preferred Latin to the vernacular.[24]

In addition to words like *Adeling, Ciricsceat, Cuningham, Cynedom, Danegelt, Gavelkind, Mundbreche*, *Socne*, which are all defined in Latin though their origin is clearly Germanic, there are entries such as *Lannemanni, Mundeburdum, Sidlingis, Taini*, whose Latin form might mislead the reader. They are, in fact, Anglo-Saxon words which entered the linguistic heritage of English law in a Latinate form. For *Lannemanni* the main meaning is given in English followed by the explanation in Latin, while *Munderburdum* is defined only in Latin and *Sidlingis* only in English:

(24) **Taini** or Thaini mediocres were Freeholders and sometymes called milites regis, and their land called Tainland. Cokes Inst: part 1 fol: et fol: 86.a.[25]

(25) **Lannemanni**.
signifyeth Lords of a Mannour Coke having Socam et Sacam de Inst: part tenentibus et hominibus suis fol. 5.a[26] Civitas Lincol. Item in ipsa civitate

24 Francis Bacon, for example, had published his *Advancement of Learning* in 1605 and then, in 1623, translated the same work into Latin, *De Dignitate et Augmentis Scientiarum*. See also fn. 11: many of Dugdale's sources were historical works written in Latin.

25 The source reads: "*Taini* or *thaini mediocres* were freeholders, and sometime called *milites regis*, and their land called *Tainland*". Coke's Institutes part I, fol. 5. b.

26 The source reads: "*Lannemanni* there signifie lords of a mannor, hauing *socam et sacam de tenentibus et hominibus suis*. Cokes Institutes part 1 fol: 5.a".

erant xii Langemanni, id est habentes socam et sacam. Lib: Domesday. Lincolsc.

(26) **Mundeburdum**.
Pippinus (ℭ)) Aquitaniae Rex (ℭ)) ut omnes res ad suprascriptum monasterium (ℭ)) sub nostro mundeburdo,[27] vel immunitatis tuitione recepimus (ℭ))Preuues de l'Hist: des Comtes de Guyene. p. 24.

(27) **Sidlingis**. Sides or bulkes, betwixt erable ridges or lands. vide Buttum[28]

Entries occur in which English and Latin are mixed, such as in *Wreccum-maris* and *Thainus regis*:

(28) **Wreccum maris**. vide rot: chart: de anno 20. H. 3 m¯. 3 et 4. quidquando sit wreccum. Et Rot: cart: 14. H. 3 6. Et Pat: 42. H. 3. m¯. 1. In dorso

(29) **Thainus regis**. is taken for a Baron. Thainus regis proximus Comiti est, etc. Cokes Inst: part 1 fol: 5.b

and with English and French, such as *Cherchesonde*, an Anglo-Saxon word recorded in a corrupted form influenced by French (as clarified in the source quoted below):

(30) **Cherchesonde**. une mesure de ble que checun homme soleit envoier a seint Esglise en temps de Bretons. Seldens Hist: of Tithes p: 216

That *Cyrycsceat* is a Church-rent of Corn, or the first fruits of Corn yeerly in those times, and regularly payable at S. *Martins* day to the Church; and is sometimes written *Cureseet*; sometimes otherwise. And in an old Ms. Exposition of Law-terms, occurres, ***Cherchesonde, vne mesure de Ble que checun homme soleit enuoier a Seint Eglise en temps de Bretons***. Plainely, Church-Corn is vnderstood; and *Cyrksceat*, that is, Church-rent is the originall whence *Cherche-sonde* is there corrupted. (Selden 1969: 216)

As a matter of fact, the glossary bears witness to all the processes of language integration and adaptation that have characterized the history of the English language – Anglo-Saxon entries co-exist with forms

---

27 Underlined in the MS.

28 **Buttum**. A butt or a ridge of erable land in Carta Mauricii de Sibbesford -- Dedi decen acras et vnum Buttum cum capuciis et sidlingis prati ad eandem terram pertinentibus.

whose ultimate origin is Celtic, Latin, French and Latinate, such as *brache, glyn, dongeon, scute*, showing how English vocabulary has been changing, increasing and constantly enriching its expressive potential. A long history of contacts between languages and cultures – from the Celtic dawn, through Roman rule, the settlement of west Germanic tribes, the Viking invasions until the Norman Conquest – is highlighted in every linguistic item.

5. Concluding remarks

If we consider Dugdale's written output as a whole, we can define it prodigious: it represents a lifetime of hard work among the surviving legal, civil, ecclesiastical and familial records. In a career of sustained research from the 1630s to the 1680s, Dugdale carried out his persistent attempt to clarify the development of medieval institutions. He did not believe that the Middle Ages were sunk in ignorance and monkish superstition. As Graham Parry (1995: 217) writes:

> He recognised that medieval society had been dominated by disciplined institutions, which had conditioned the growth of the nation, and that he and his contemporaries lived in a world that had been shaped by the social structures that had evolved in a time of feudalism and universal faith.

His lexicographical works, unfortunately mostly unknown, emphasise how Dugdale retrieved information in order to recover factual matter relating to the development of great institutions during the Middle Ages such as the legal system, the aristocracy and the monasteries. The greatest obstacle was, indeed, the disorder of the records he worked on and drew information from. They lay in utmost confusion and often in "mouldering neglect" (Parry 1995: 225); they were scattered during and after the Reformation. Family records had to be solicited from interested parties; churches all over the country had to be constantly visited and their inscriptions and coats of arms painstakingly transcribed.

What emerges from the analysis of Dugdale's lexicons and, in particular, of this unpublished and unfinished glossary, is the antiquarian's method of making searches into manuscript evidence, interpreting it and making comparisons for authenticity (Parry 1995: 246). The study of charters and the knowledge of ancient laws were crucial to carry out this kind of activity. Thanks to Spelman, who trained Dugdale to learn Anglo-Saxon, and Somner, who offered him enormous assistance in this field, Dugdale realised the importance and the need to know about Anglo-Saxon for antiquarian research in the legal and institutional fields. MS Harley 1129 is the mirror of specialised and unspecialised legal terminology; it is the expression of an English world which is a melting pot of languages and cultures.

References

Primary sources

London, British Library, MS Harley 1129.

Oxford, Bodleian Library, MS Dugdale 29.

Boece, Hector 1527. *Scotorum Historiae a prima gentis origine.* Amsterdam: Theatrum Orbis Terrarum 1977.

Coke, Lord Edward 1628. *The Institutes of the Laws of England. Or a Commentarie upon Littleton, not the name of a Lawyer onely, but of the Law itselfe.* [...] *Authore Edward Coke Milite*, Printed for the Societie of Stationers, London: Hargraves [13]1832.

Cotgrave, Randle 1611. *A Dictionarie of the French and English Tongues.* New York: Hildesheim 1970; reprint of London: Islip.

de Hoveden, Rogerus *Chronica Magistri Rogeri Hovedene edited by William Stubbs.* Millwood: Kraus Reprint 1964. See also www.gallica.bnf.fr/scripts/ConsultationTout.exe?O=N050282.

Preuves de l'Histoire des Comtes de Poictou et Ducs de Guyenne. In Besly, Jean 1647. *Histoire des Comtes de Poictou et Ducs de Guyenne* Parigi: R. Bertault.

Selden, John 1618. *The Historie of Tithes*. Amsterdam: Theatrum Orbis Terrarum; New York: Da Capo Press, 1969.

Secondary sources

Berkhout, Carl T. / Gatch, Milton McC. (eds) 1982. *Anglo-Saxon Scholarship. The First Three Centuries*. Boston (Mass.): Hall.

Darby, Henry C. 1977. *Domesday England*. Cambridge: Cambridge University Press.

Du Cange, Charles Du Fresne 1883-1887. *Glossarium mediae et infimae Latinitatis*. Niort: Favre.

Dyer, Christopher 1985. Towns and Cottages in Eleventh-century England. In Mayr-Harting, Henry / Moore, Robert Ian (eds) *Studies in Medieval History presented to Ralph Henry Carless Davis*. London: Hambledon, 91-106.

Dyer, Christopher 1994. *Everyday Life in Medieval England*. London: Hambledon.

Hamper, William (ed.) 1827. *The Life, Diary and Correspondence of Sir William Dugdale*. London: Harding, Lepard.

Hetherington, Sue M. 1980. *The Beginnings of Old English Lexicography*. Spicewood, Texas: unpublished typescript.

Lennard, Reginald Vivian 1959. *Rural England, 1086-1135: A Study of Social and Agrarian Conditions*. Oxford: Clarendon.

Maitland, Frederic William 1897, repr. 1989. *Domesday Book and Beyond*. Cambridge: Cambridge University Press.

Nares, Robert 1808-1812. *A Catalogue of the Harleian Manuscripts in the British Museum with Indexes of Persons, Places and Matters*. 4 vols. London: Eyre & Strahan.

Parry, Graham 1995. *The Trophies of Time. English Antiquarians of the Seventeenth Century*. Oxford: Oxford University Press.

Tornaghi, Paola 1999. Sir William Dugdale e il *Dictionarium Saxonico-Anglicum*. *Linguistica e Filologia* 10: 141-166.

Walters, Henry Beauchamp 1934. *The English Antiquaries of the Sixteenth, Seventeenth and Eighteenth Centuries*. London: Walters.

Appendix

Socne.[29] id est reatus emendatio. Reg: de Thorney penes mildmay Comit. westmert. fol: 4.a

Ex libro irrotulamentorum, extens diuersorum maneriorum, et aliarum terrarum monasterium ecclesiae Christi Cantuar. tangentium fol: 211

Milites seu liberi tenentes de feodo militari debent	esse in custodia usque ad xxj annos. homagium facere relevare terram. maritari per dominum. dotari de tertio. socagium praestare. primogenitus succedit in toto
Gavelkind	debent feodalitatem facere. esse in nutritura propinquioria consanguinei cui haereditas post-seipsum non pertinet, usque ad xv annos recognicionem domino facere pro terra sua. dotari de medio. omnes participant
Libera Sokemanzia	possunt dare aut vendere, sed ad voluntatem domini. non alienare certa seruicia. antenatus succedit in totum. Averium masculum non vendere. filiam non maritare, nisi dat domino iii^{s}. iiii^{d}. filium omnino facere clericum et isti dicuntur Sokemanni liberi.
Sokemanzia quae dicitur Coteria	debent talliari ad voluntatem domini. facere servicia incerta nihil dare, nihil vendere. nihil proprium habere, nec potest acquirere, nisi ad promotionem domini sui. Et isti dicuntur Cotarii

29 The term *soke* (OE *soc*) at the time of the Norman Conquest generally denoted 'jurisdiction', but due to vague usage probably lacked a single precise definition. In some cases *soke* denoted the right to hold a court, and in others only the right to receive the fines and forfeitures of the men over whom it was granted when they had been condemned in a court of competent jurisdiction.

HANS LANDQVIST

Continuity and Change: The Swedish Legal Lexicon as Seen from the Perspective of a Monolingual Specialist Dictionary

1. Introduction

Like the general lexicon, a legal lexicon in a particular language evolves over time (cf. Aitchison 2001: 4). Exploration of monolingual legal dictionaries is one way to study continuity and change in the legal lexicon of a language. To my knowledge this avenue has been parsimoniously used, to say the least, in earlier studies of the legal lexicon in various languages (cf. Tiersma 1999; Mattila 2002). There are essentially no previous studies of the Swedish legal lexicon from a diachronic perspective based on monolingual specialist dictionaries (see Landqvist 2003: 9-10).

In an earlier study of the Swedish legal lexicon in 1800-2000 (Landqvist 2003), I worked with a subset of material based on two general dictionaries and a subset based on four monolingual legal dictionaries. I studied loanwords and native word-formation based on material comprising a total of 2,003 headwords from *Svenska Akademiens ordbok* (1893-, 'Dictionary of the Swedish Academy') and *Nationalencyklopediens ordbok* (1995-1996, 'Dictionary of the National Encyclopaedia'). Certain results derived from the general material were compared with results derived from the specialized material of 10,811 headwords compiled from four monolingual legal dictionaries published after 1950: *Juridikens begrepp* (2000, 'Legal Concepts'), *Juridikens termer* (1958, 'Legal Terminology'), *Juridikens termer* (1992, 'Legal Terminology'), and *Norstedts juridiska ordbok* (1998, 'Norstedt's Legal Dictionary'). I also compared certain headwords within the subset of specialized material. It emerged that approximately 40% of the headwords appeared in only one of the specialized dictionaries and slightly more than 31% in two of them, while 17% of the headwords were included in three of the dictionaries and slightly more than 11% in all four (see Landqvist 2003: 109-110).

The study presented here is based on those results, having observed the absence of previous diachronic investigations of the Swedish legal lexicon based on specialized dictionaries. The aim is to examine the changes that have occurred during a relatively short period of time (the past fifty years) by making use of two editions of the same legal dictionary. Although specific reference is made to the Swedish legal lexicon, its methodology may of course be employed for diachronic studies of the legal lexicon in other languages – for instance, different editions of *Black's Law Dictionary* ([8]2004) would be a likely source of Anglo-American Legal English material. In the case in point, my analysis is based on two editions of the monolingual Swedish legal dictionary *Juridikens termer* ('Legal Terminology',[1] henceforth *JuT*). The first edition was published in 1958 and the ninth and most recent one in 2002 (Eek *et al.* 1958; Bergström *et al.* 2002). The headwords that appear in both editions may be said to indicate continuity in the Swedish legal lexicon, while those that appear in only one would appear to indicate change. The results of the study are discussed from the perspective of linguistic and social explanations (cf. Aitchison 2001: 133-183).

2. Material and points of departure

The material studied consists of the lemmas that appear in the two editions of *JuT*. Like Svensén (2004: 115), I use 'lemma' to refer to

1 The study may be regarded as a development of my previous research project on the Swedish legal vocabulary in the nineteenth and twentieth centuries (Landqvist 2003), a sub-project of the ORDAT project (Sw. *Det svenska ordförrådets utveckling från artonhundra till tjugohundra*; Eng. 'Developments in Swedish Vocabulary from 1800 to 2000'). A formal presentation of the ORDAT project is given (in Swedish) in Malmgren (2000).

the dictionary entries in bold type that consist of one or more text words. Analysis of the lemmas was performed in part based on the conceptual system and various semantic relationships between legally regulated concepts (Section 3.2) and in part using five contextual categories, which are presented in Section 3.3.

JuT usually gives only one meaning for each lemma, but in some cases provides explanations of adjacent lemmas under a particular lemma. For instance, the three terms FRUCTUS CIVILIS (fructus civiles / *civil fruits*), FRUCTUS NATURALIS (fructus naturales / *natural fruits*) and FRUCTUS PERCIPIENDI (perceived fruits / *the customary produce of property*) are explained under the superordinate term FRUCTUS (fructus / *fruits or income*).[2] This is thus a question of a hyponymy / hyperonymy relationship (cf. Sandrini 1996: 100; Pilke 2000: 30-36). In a few cases, multiple meanings, i.e. instances of polysemy, are given for a single lemma (cf. Pilke 2000: 30). Thus, FULLMAKT (power of attorney) has one meaning in Swedish civil law and another in constitutional law. In a likewise limited number of cases, synonyms or near-synonyms are given, e.g. FOLKMORD (GENOCIDIUM) (genocide) (cf. Sandrini 1996: 253-254; Tiersma 1999: 113).

The points of departure for the present study are the common lemmas and unique lemmas in the material, i.e., the lemmas found in both editions of *JuT* versus those found in only one of them. From this point on, the editions will be referred to as *JuT1958* and *JuT2002* respectively. As the lemmas are included in the respective editions, the editors obviously found they were part of a Swedish legal lexicon in 1958 and 2002. Despite that, the lemmas are of discrete nature, as legal jargon does not use only defined terms, but also the resources of the general lexicon (see Tiersma 1999: 106-110). For that reason, the material includes defined single-word and multiple-word terms, words and phrases from the general lexicon that have particular legal meanings, as well as acronyms and proper nouns. Lemmas that give

---

2 Lemmas are shown in UPPER CASE, followed by English 'legal Latin' equivalents where relevant, and approximate English translations in parentheses, in lower case. Translations of definitions are in *italics*. The nature of law as a national field of knowledge is such that the English translations of lemmas and definitions are not always the exact legal equivalents (see Bergenholtz / Tarp 1994: 57-58).

only references to other lemmas in each edition are also included (cf. Svensén 2004: 115, 461). Examples of types of lemmas under the letter S are shown in Figure 1.

Type of lemma	*Example*
Single- and multiple-word terms	SAKÄGARE 'subjekt i rättsförhållande' (*JuT1958*; *JuT2002*) (person concerned / interested party / *subject in a legal relationship*) STRICTO JURE 'strängt efter rätten' (*JuT1958*; *JuT2002*) (stricti juris / *of the strictest right or law; according to the exact law*)
Words and phrases from the general lexicon with particular legal meanings	SKEPP 'fartyg vars skrov är minst tolv meter långt och minst fyra meter brett, medan annat fartyg enligt sjölagen kallas båt' (*JuT2002*) (ship / *vessel whose hull is at least twelve meters long and at least four meters wide, while other vessels are pursuant to the Swedish Maritime Act referred to as boats*) SUMMARISK LÄKARUNDERSÖKNING 'mindre omfattande läkarundersökning av den tilltalade i ett brottmål, vilken vanligen utförs av en psykiater' (*JuT1958*) (abridged medical examination / *less extensive medical examination of the accused in a criminal case, usually performed by a psychiatrist*)
Acronyms	SSNIP-test (small but significant non-transitory increase in prices) 'metod för att bestämma en varas utbytbarhet' (*JuT2002*) (SSNIP test / *method of determining the substitutability of goods*)
Proper nouns	SÄKERHETSRÅDET 'FN:s säkerhetsråd' (*JuT2002*) (The UN Security Council)
Reference lemmas	STRAFFVITE 'se VITE' (*JuT1958*; *JuT2002*) (punitive fine / *see* fine)

Figure 1. Examples of types of lemmas in the material.

3. Results and discussion

I report and discuss my results in three sections. The first deals with the lemmas in the material, followed by the lemmas common to both editions and finally the unique lemmas. I conclude with a discussion of the results from a more general perspective.

3.1. Lemmas

The number of lemmas in the material is summarized in Table 1.

Lemmas	*JuT1958*	*JuT2002*
Common lemmas	2,370	2,370
Unique lemmas	578	982
Total	2,948	3,352

Table 1. Lemmas in *JuT1958* and *JuT2002* (absolute numbers).

The material includes a total of 6,300 lemmas including 2,948 from *JuT1958* and 3,352 from *JuT2002*. This does not mean that there are 6,300 different lemmas in the material, as 2,370 lemmas appear in both editions. One possible explanation of the many common lemmas may be that the central Swedish legal lexicon did not change appreciably between 1958 and 2002 (cf. Tiersma 1999: 87). Another possible explanation is that the editors of later editions of *JuT* relied heavily on the work done by the editors of *JuT1958* (cf. Svensén 2004: 50). In addition to common lemmas, the material contains a total of 1,560 unique lemmas, 578 from *JuT1958* and 982 from *JuT2002*. The quantitative difference between the two editions with regard to the number of unique lemmas may be due simply to greater ambition on the part of the editors (cf. Bergström *et al.* 2002: 3). Another explanation may be that Swedish society became increasingly complex and regulated in the nearly fifty years between the two editions, leading the editors of *JuT2002* to judge that more lemmas should be included. I was unable to find any earlier studies with the same orientation as this one (see Section 1). Consequently, it is difficult to determine whether the distribution of common and unique lemmas in the material is noteworthy or not.

3.1.1. Common lemmas

We can presume that the 2,370 common lemmas in the material form a kind of core in the Swedish legal lexicon current in the latter half of the twentieth century. A traditional perspective on the lexicon of a particular language at a particular point in time is to study native words and loanwords respective to native word-formation with patterns of compounding, derivation and blending (see Edlund / Hene 1992: 28-32; Aitchison 2001: 14; Aitchison 2003: 127). However, a perspective more oriented towards terminology is more fruitful here, as the editors of *JuT* seem, albeit not exclusively, to have selected the lemmas based on terminological principles rather than lexicological or lexicographical ones (cf. Sandrini 1996: 195; Pilke 2000: 32-34). The selection is thus based primarily on legal concepts and associated terms, while general dictionaries are usually based on words and possibly some jargon terms (cf. Svensén 2004: 90, 299).

A simplified description of the legal system in a particular society is that it is meant to regulate rights and obligations among various actors in a variety of situations. For that reason, rules of law are devised which form a legal system. Various actors interpret, monitor and regulate relationships and general human behavior in society. Conflicts are regulated, often using sanctions, the ultimate aim of which is to maintain peace and order (see Hydén 2001: 9-24, 37-38). Conceptual systems with varying relationships among legally regulated concepts become central in this context (Pilke 2000: 58-60).

The argument may be exemplified based on the situation in which two of the central actors are designated BORGENÄR (creditor) and GÄLDENÄR (debtor). A third actor, the BORGEN(S)MAN [also BORGESMAN or LÖFTESMAN] (guarantor or surety), assumes liability for that the debtor will pay his debt. The guarantor provides a BORGEN (surety or guarantee). The surety may take various forms: ENKEL BORGEN (simple guarantee), PROPRIEBORGEN, (guarantee with direct liability or personal surety), UNDERBORGEN (underlying guarantee / surety), and ÖVERBORGEN [also EFTER-BORGEN] (guarantee / surety for a guarantee). The situation may play out in a variety of ways depending on how the actors assert their rights and fulfill their obligations (see Håstad 1994: 204-207).

The conceptual relationships may be classified in various ways from a terminological perspective (Pilke 2000: 58-60). The hopefully clearly defined legal concepts and their related terms may also be said to build a kind of semantic network (cf. Aitchison 2003: 84-101), within which various semantic relationships can be identified. In this case, we find converseness (*borgenären-gäldenären / creditor-debtor*), hyperonymy (*borgen-enkel borgen / guarantee-simple guarantee*) and hyponymy (*enkel borgen-borgen / simple guarantee-guarantee*). Synonymy is avoided, on the other hand, as the ideal in a specialized field is a one-to-one relationship between concept and term (see Sandrini 1996: 253; Pilke 2000: 35, 284). This, however, is not entirely successful, as *borgen(s)man*, *borgesman* and *löftesman* and *överborgen* and *efterborgen* are classified as synonymous designations in *JuT1958* and *JuT2002*.

3.1.2. Unique lemmas

As seen in Table 1 above, the number of unique lemmas is greater in *JuT2002*, with 982 items compared to the 578 found in *JuT1958*.

There are probably several reasons why a particular lemma is found in only one edition (cf. Aitchison 2001: 134). Firstly, this may depend on the editors' selection of more or less exclusively legal terms. As an example, ASYLRÄTT (asylum right / *the right of a state to protect a refugee from the authorities in his country of origin without the measure being regarded as a hostile action*) is unique to *JuT1958*, while ASYL (asylum) is found only in *JuT2002*. It may also depend on the selection of words and phrases from general language with particular legal meanings (Pilke 2000: 30, 285). The penal law term FYLLERI (public drunkenness or intoxication) is found only in *JuT1958*, while the commercial law term BEFINTLIGT SKICK (as-is condition) is unique to *JuT2002*. The editors further make a selection of acronyms and proper nouns, such as MBL (MEDBESTÄMM-ANDELAGEN) (*JuT2002*) (The Swedish Co-Determination at Work Act) and MILITIEOMBUDSMANNEN (*JuT1958*) (Parliamentary Commissioner for Military Affairs). Secondly, many of the unique lemmas in each edition were formed in response to a need brought about by changes in Swedish national law or international law (cf. Aitchison 2001: 146). In an attempt to clarify some causes, I have sorted the unique lemmas into five contextual categories, which may be said to roughly correspond to branches of law (cf. Hydén 2001: 25). The classifications were based on the information given about the lemmas in the editions. In alphabetical order, the five categories are:

1. European Union,
2. International legal relationships,
3. IT and associated concepts,
4. Roman law,
5. Swedish institutions and persons with legal functions and Swedish legally defined concepts.

I will first give my reasons for the choice of categories and thereafter present and discuss my results. I chose the three contextual categories 'European Union,' 'International legal relationships,' and 'IT and associated concepts' on account of the evolution of law in the latter twentieth century towards an increasingly globalized world, one in which information technology is playing a significant role (Hydén 2001: 55-82, 45-51). Lemmas include the EC law term ACQUIS COMMUNAUTAIRE (*JuT2002*) (*the entire body of EC legislation*), the treaty law term RATIFIKATION (*JuT1958*) (ratification) and the IT term ELEKTRONISKT DOKUMENT (*JuT2002*) (electronic document).

I chose the category 'Roman law' because Roman law – and Latin – may, as a result of European integration, be more significant to the current Swedish legal lexicon than it has been for centuries (cf. Mattila 2000: 305). The category includes both lemmas based directly on Roman law and lemmas that entered the Swedish legal lexicon via other legal traditions such as the Germanic and the Anglo-American ones (see Mattila 2002: 201-266). Examples include DARE (*JuT1958*) (*to give*) and EX GRATIA (*JuT2002*) (also a gratia, *by favor*).

The category 'Swedish institutions and persons with legal functions and Swedish legally defined concepts' differs from the other four in that it is broader in nature. In contrast to the other four, it is also focused on national conditions, as law, unlike, for instance, medicine and technology, may be designated as a culturally determined area of knowledge (see Bergenholtz / Tarp 1994: 57-58). Lemmas in this category include ALLMÄNHETENS PRESS-OMBUDSMAN (PO) (*JuT2002*) (Public Press Ombudsman), CHEF-RÅDMAN (*JuT2002*) (Senior Judge) and DRÅPSBROTT (*JuT1958*) (manslaughter).

Instances of lemmas unique to *JuT1958* and *JuT2002* based on the five contextual categories are reported in Table 2.

Contextual category	*JuT1958*	*JuT2002*
European Union	–	86 (8.8)
International legal relationships	20 (3.5)	76 (7.7)
IT and associated concepts	–	2 (0.2)
Roman law	11 (1.9)	23 (2.3)
Swedish institutions and persons with legal functions and Swedish legally defined concepts	547 (94.6)	795 (81.0)
Total	578 (100.0)	982 (100.0)

Table 2. Lemmas unique to *JuT1958* and *JuT2002* sorted into contextual categories (in absolute numbers and percentages).

As I see it, the unique lemmas in the two editions exist partially due to the need for them that had arisen according to the respective editors at the time, consequent upon developments in Swedish or international law. What does the information in the table tell us about those developments? Relying on Table 2, one may assert that the lexicon Swedish lawyers are thought to need became more international from the late 1950s to the early 2000s. Firstly, there are 86 unique lemmas for the 'European Union' category in *JuT2002*, including EUROPA-PARLAMENTET (The European Parliament) and FÖRHANDS-AVGÖRANDE (preliminary ruling *by the European Court of Justice*). Secondly, *JuT1958* is notable for only 20 unique lemmas related to 'international legal relationships' while that category accounts for 76 instances in *JuT2002*. Examples include SUZERÄN (suzerain) which is, according to *Juridikens termer*, an alternative term for VASALL-STAT (*JuT1958*) (vassal state).[3] Another is STATUT (*JuT2002*)

3 As *Black's Law Dictionary* (2004), defines *suzerain* as "A nation that exercises control over another nation's foreign relations," i.e., the opposite of a vassal state, it is somewhat perplexing that *JuT1958* refers to SUZERÄN and VASALLSTAT as synonymous.

(statute), which is used in international private law concerning applicable law.

There are only two instances for the category of 'IT and associated concepts': ELEKTRONISK ANSLAGSTAVLA (electronic bulletin board) and ELEKTRONISKT DOKUMENT (electronic document), both found in *JuT2002*. This is somewhat surprising, as there are 28 instances for the category in the monolingual legal dictionary *Juridikens begrepp* ('Legal Concepts'). The first edition was published in 2000 and contains 2,869 lemmas (see Landqvist 2003: 30). *Juridikens begrepp* is touted as containing "everything from a number of relevant terms from classic Roman law to the latest in EC law and IT law" (Melin 2000: 5).

There are a total of 34 unique lemmas for the 'Roman law' category in the material, accounting for about the same percentage in both editions, at 1.9% for *JuT1958* and 2.3% for *JuT2002*. Examples include the international public law term OCCUPATIO BELLICA (*JuT1958*) (*belligerent or military occupation*) and the maxim of penal law NULLA POENA SINE CRIMINE (*JuT2002*) (*no punishment without a crime*). A review of the material shows that it also contains 217 lemmas that are common to both editions of *Juridikens termer* and are Latinisms, i.e., loanwords or phrases from Latin that retain their original form and spelling and act as symbols for the relevant concept belonging to a particular legal system (cf. Mattila 2002: 517). The unique lemmas together with the common Latinisms may be interpreted to mean that Swedish lawyers still require some understanding of Roman law and of Latin, partly as a consequence of Swedish membership in the European Union (cf. Mattila 2000: 305). Explicit information that such lemmas are related to the European Union is however given for only three unique lemmas in *JuT2002*. One such is FUMUS BONI JURIS ('*the smoke of good law*'), which is used to denote that a precedent exists for the European Court of Justice to grant an application for an interim measure.

Table 2 shows that one contextual category is predominant among the unique lemmas: 'Swedish institutions and persons with legal functions and Swedish legally defined concepts'. The category accounts for nearly 95% of the unique lemmas in *JuT1958* and slightly more than 80% in *JuT2002*. The result is in part due to the scope of the category compared to the other four, but it should still be reasonable to claim that it illustrates the nature of law as a culturally determined field of knowledge. The lemmas put in this category may in turn be broken down into three subcategories: 'newly introduced institutions, persons and concepts'; 'existing institutions, persons and concepts given new designations' and 'institutions, persons and concepts that no longer exist in the legal sense'.

I will illustrate the first subcategory of 'newly introduced institutions, persons and concepts' based on the emergence of Swedish environmental law. Environmental issues became increasingly topical in the latter half of the twentieth century and a few related lemmas unique to *JuT2002* are MILJÖBALKEN (The Swedish Environmental Code), MILJÖBROTT (environmental crime), MILJÖDOMSTOL (Environmental Court) and MILJÖ-PROCESS (environmental process). The Rixlex database of Swedish statutes shows that the Swedish Environmental Damage Act entered into force in 1986 and the Swedish Environmental Code in 1998, meaning that the relevant lemmas should have been established or lexicalized in Swedish during the past twenty years. However, when does a lemma actually become established in the field of law?

It seems reasonable from the linguistic perspective to regard a particular lemma as established or lexicalized once the preliminary documents (such as government bills, parliamentary motions or official reports) upon which statutory texts are based exist or as soon as a statutory text has been issued by the Swedish Riksdag (laws) or the Government (ordinances). But the question is whether any of these situations is in fact adequate from the legal standpoint or whether a statute must have taken legal force for lawyers to regard a particular lemma as established or lexicalized (see Landqvist 2003: 121).

The subcategory of 'existing institutions, persons and concepts given new designations' is exemplified by the lemma FATTIGVÅRD (poor relief) in *JuT1958*. FATTIGVÅRD is defined as "*a general form of aid provided by municipalities that is regulated by various laws dating from the latter nineteenth century and thereafter in the 1918 act on poor relief in the municipalities*". *JuT1958* also informs the reader that the 1918 act on poor relief had been superseded by the 1956 Social Assistance Act; in turn, this was replaced by the Social Services Act, which was issued in 1980 and took effect two years later (see Inger 1988: 297). SOCIALHJÄLP (social assistance) is found in

JuT1958, but not in *JuT2002*, while SOCIALTJÄNST (social services) appears only in *JuT2002*. Through legislative reform, the concept of 'fattigvård' (*poor relief*) in the legal sense ceased to exist and was replaced by other terms: first 'socialvård' (*social care*) and later 'socialtjänst' (*social services*) (cf. Tiersma 1999: 98).

LÖSDRIVARE (vagrant) is an example taken from the sub-category of 'institutions, persons and concepts that no longer exist in the legal sense.' A lemma unique to *JuT1958*, it is defined as "a person who lacks means of support and also either moves idly from one place to another or also fails to honestly support himself and moreover pursues a manner of living that is a threat to public safety, order, or morality [...]". LÖSDRIVARE is thus not included in *JuT2002* and a search of the concordance material in Språkbanken (The Bank of Swedish – A Linguistic Reference Data-base of Göteborg University) returns 40 hits for *lösdrivar**, most originating in novels. The word *lösdrivare* thus does not seem to exist as a native Swedish legal term in the latter half of the twentieth century. A search of the Rixlex database does however return one hit for *lösdrivare* in the 1994 law that governs the validity of the European Convention for the Protection of Human Rights and Fundamental Freedoms. In that text, *lösdrivare* is used as the Swedish translation of *vagrant* and *vagabond* in the English and French versions of the Convention. This implies that even in international legal contexts *lösdrivare* was rarely used as a Swedish legal term in the late twentieth century.

4. Concluding discussion

In the present study, I have explored the lemmas in the first edition of the monolingual Swedish legal dictionary *Juridikens termer* ('Legal Terminology') from 1958 and the ninth and latest edition from 2002. The total number of lemmas was 6,300, including 2,948 from the 1958 edition and 3,352 from the 2002 edition. All told, 2,370 lemmas are common to both editions, which seems to indicate relatively great continuity in the Swedish legal lexicon of the second half of the twentieth century. There is also a measure of change through the lemmas found in only one of the editions, with 578 unique lemmas in the 1958 edition and 982 in the 2002 edition.

A sub-study of the unique lemmas in both editions shows that the lexicon Swedish lawyers are judged to need may be said to have become more international in the intervening years from the late 1950s to the early 2000s. On the other hand, developments in society at large seem to have left only negligible traces in the material with regard to information technology. The unique lemmas stemming from Roman law, when combined with the Latinisms found in both editions, may be interpreted to mean that Swedish lawyers are still thought to need some understanding of Roman law and of Latin. Most of the unique lemmas originate in Swedish law, which illustrates the nature of law as a culturally determined field of knowledge. The tendency towards change in a Swedish legal lexicon during the period studied is also evident in the lemmas within the category of 'Swedish institutions and persons with legal functions and Swedish legally defined concepts' that are introduced, given new designations or have ceased to exist in the legal sense.

The question is: how reliable are the results with respect to the lexicon that professionals in the field need and understand? Owing to various practical circumstances that affect the production of dictionaries (source material, staff, technical tools, etc.) as well as the matter of establishment or lexicalization of the lemmas, or the meanings of lemmas, no dictionary can ever perfectly reflect the lexicon of a particular language at a particular point in time (cf. Pilke 2000: 33 and Aitchison 2003: 11-14). A familiar description of the establishment or lexicalization of lemmas is given by Samuel Johnson in the preface to his *Dictionary of the English Language* (1755): "No dictionary of a living tongue can ever be perfect, since while it is hastening to publication, some words are budding, and some fading away" (quoted in Aitchison 2003: 11). It follows that the results presented above do not provide the only picture of the lexicon that Swedish legal professionals needed or mastered in the late 1950s or the early 2000s, but they nevertheless show tendencies in the evolution of such a lexicon.

The results of this pilot study show however that monolingual specialist dictionaries are useful for the exploration of aspects of lexical development in specialized fields of knowledge (cf. Landqvist 2003: 28-30). Another avenue is to use diachronic general dictionaries to create corpora based on professional terminology and lemmas in such dictionaries related to specialized fields (see Landqvist 2003: 15-28). Further opportunities for diachronic studies of lexical development in various fields may derive from corpora based on literature in the field, both that intended for professionals and that intended for interested laymen (cf. Landqvist 2003: 24).

References

Primary sources

Bergström, Sture / Andersson, Torbjörn / Håstad, Torgny / Lindblom, Per Henrik [9]2002. *Juridikens termer.* [Legal Terminology]. Stockholm: Liber.

Bergström, Sture / Håstad, Torgny / Lindblom, Per Henrik / Rylander, Staffan [8]1992. *Juridikens termer* [Legal Terminology]. Stockholm: Almqvist & Wiksell.

Black, Henry Campbell [8]2004. *Black's Law Dictionary.* Garner, Bryan A. (Editor in Chief). St. Paul MN: West Publishing Co.

Eek, Hilding / Bergström, Svante / Arnheim, Frans 1958. *Juridikens termer. Presenterade av Hilding Eek Svante Bergström Frans Arnheim* [Legal Terminology. Presented by Hilding Eek, Svante Bergström, Frans Arnheim]. Stockholm: Svenska bokförlaget.

Martinger, Sven [4]1998. *Norstedts juridiska ordbok. Juridik från A till Ö* [Norstedt's Legal Dictionary. Law from A to Z]. Stockholm: Norstedts Juridik.

Melin, Stefan 2000. *Juridikens begrepp* [Legal Concepts]. Uppsala: Iustus.

Nationalencyklopediens ordbok. Utarbetad vid Språkdata Göteborgs universitet 1995-1996 [Dictionary of the National Encyclopaedia. Produced by the Sections of Natural Language Processing and Lexicology, Department of Swedish, Göteborg University]. Höganäs: Bokförlaget Bra Böcker.

Språkbanken Konkordanser i fulltext [The Bank of Swedish Full-text Concordances], Göteborg University <spraakbanken.gu.se/lb/konk>, accessed July 2005.

Svenska Akademiens ordbok 1893- [The Dictionary of the Swedish Academy]. Lund: Gleerupska Universitetsbokhandeln.

Swedish Riksdag. Rixlex Författningar i fulltext [Full-text Swedish Statutes], <www.riksdagen.se/debatt/sfst/index.asp>, accessed July 2005.

Secondary sources

Aitchison, Jean ^{3}2001. *Language Change: Progress or Decay?* Cambridge: Cambridge University Press.

Aitchison, Jean ^{3}2003. *Words in the Mind. An Introduction to the Mental Lexicon*. Oxford: Blackwell.

Bergenholtz, Henning / Tarp, Sven 1994. *Manual i fagleksikografi. Udarbejdelse af fagordbøger – problemer og løsningsforslag* [Manual of Specialized Lexicography: The Preparation of Specialized Dictionaries]. Herning: Systime.

Edlund, Lars-Erik / Hene, Birgitta 1992. *Lånord i svenskan. Om språkförändringar i tid och rum* [Loanwords in Swedish. On Linguistic Change in Time and Space]. Höganäs: Wiken.

Håstad, Torgny 1994. Andra särskilda avtal [Other Special Contracts]. In Sandgren, Claes (ed.) *Norstedts juridiska handbok* [Norstedt's Legal Handbook]. Stockholm: Fritzes, 171-210.

Hydén, Håkan ^{5}2001. *Rättsregler. En introduktion till juridiken* [Rules of Law. An Introduction to Law]. Lund: Studentlitteratur.

Inger, Göran ^{2}1988. *Svensk rättshistoria* [Swedish Legal History]. Stockholm: Liber.

Landqvist, Hans 2003. *Abolition, militärmål, rusdryck och kåkfarare. Studier i ett svenskt juridiskt ordförråd 1800-2000* [Abolition, Military Case, Strong Drink, and Jailbirds. Studies in a Swedish Legal Lexicon 1800-2000]. ORDAT 20. Göteborg: Göteborgs universitet.

Malmgren, Sven-Göran 2000. *Projektet Det svenska ordförrådets utveckling 1800-2000. Utgångspunkter* [Research Project 'Developments in Swedish Vocabulary 1800-2000'. Points of departure]. ORDAT 1. Göteborg: Göteborgs universitet.

Mattila, Heikki E.S. 2000. Latinet i den finländska juridiska litteraturen [Latin in Finnish Legal Literature]. *Tidskrift, utgiven av Juridiska Föreningen i Finland* 156, 269-322.

Mattila, Heikki E.S. 2002. *Vertaileva oikeuslingvistiikka* [Comparative Legal Linguistics]. Helsinki: Kauppakaari. Lakimiesliiton Kustannus.

Pilke, Nina 2000. *Dynamiska fackbegrepp. Att strukturera vetande om handlingar och händelser inom teknik, medicin och juridik* [Dynamic Specialist Concepts. Structuring Knowledge Concering Actions and Events within the Fields on Technology, Medicine and Law]. Vaasa: Universitas Wasaensis.

Sandrini, Peter 1996. *Terminologiearbeit im Recht. Deskriptiver begriffsorientierter Ansatz vom Standpunkt des Übersetzers* [Legal Terminology. A Descriptive Onomasiological Approach]. Wien: Internationales Institut für Terminologieforschung.

Svensén, Bo ^{2}2004. *Handbok i lexikografi. Ordböcker och ordboksarbete i teori och praktik* [Practical Lexicography: Principles and Methods of Dictionary-making]. Stockholm: Norstedts Akademiska Förlag.

Tiersma, Peter M. 1999. *Legal Language*. Chicago: Univ. of Chicago Press.

JOANNA BUGAJ

Sources of Discourse-specific Vocabulary in Middle Scots Administrative Records: An Etymological Study of Binominals[1]

1. Legal discourse and text-types

Discourse is a cover-term for a communicative situation, taking place through a text. The text-type at the centre of my analysis is an administrative text, exemplified by materials from burgh records and acts of Parliament of early sixteenth-century Scotland. Legal texts constitute a peculiar body of data in historical linguistic studies as they are easily localisable (hence their value for historical dialectology), but their composition complies with the sophisticated requirements of the text-type itself. It is therefore necessary to bear in mind the use of conventions and formulae, as well as a "general degree of conservatism in legalese" (Meurman-Solin 2004: 31). The peculiar features of legal discourse include institutionalisation, specific rhetorics, specific lexical choices and syntactic constructions (Goodrich 1990; Tiersma 1999), where formality "presupposes stylistic choice" (Görlach 1998: 59).[2] Text-type specific, or genre-specific features[3] of administrative texts are therefore manifested not

---

1 I am grateful to the Organisers and participants of the workshop 'Diachronic Perspectives on Domain-Specific English' at the 15th European Symposium on Language for Special Purposes (29.08-2.09.2005, Bergamo, Italy) for their valuable comments and suggestions for my project.

2 On various aspects of legal discourse see, among others, Gibbons (1994), Görlach (1998), Rissanen (1999), Del Lungo Camiciotti (2004), and Włodarczyk (2005).

3 For the discussion of different approaches to the labels of *genre* and *text type* see Diller (2001), who also stresses the importance of genres in text typology and in a historical approach to LSP studies.

only on the level of syntax, which is probably their most conspicuous characteristic, but also through lexis. The purpose of a text determines vocabulary choices, and, at the same time, a particular text-type is characterised by a specific lexicon. Legal and official documents exhibit a lexical range which distinguishes them not only from everyday language but also from other specialised discourses.

The aim of my contribution is to investigate the role of the etymological background of specifically selected vocabulary in an LSP context of administrative records. This approach seeks to expand the "thesaurus framework [...] interested in [...] legal terminology" of Older Scots (Meurman-Solin 2003: 192). To illustrate etymological patterns, I have chosen binominals, a special group of lexical constructions characteristic of legal discourse, bearing in mind "the relevance of an association or correlation pattern between specific linguistic features and genre" (Meurman-Solin 2001: 8). The period under investigation is Middle Scots and the data come from the first half of the sixteenth century. As a matter of fact, Görlach (2002: 57) considers burgh records more resistant to English influence than other text types, which may be due to the fixedness and institutionalisation of such documents. This is certainly a helpful comment, as the data obtained, especially prior to 1550, can be treated as representative of Scots usage before anglicising trends became more marked (see, for instance, Devitt 1989; Agutter 1990; McClure 1994; and Jones 1997).

My study is based on two sources: corpus material from the Helsinki Corpus of Older Scots (HCOS, Meurman-Solin 1993) and an unedited MS of burgh court records from Wigtown (1512-1534: see Bugaj 2004). Choosing administrative records as the textual material for my study, I followed the genre-categorisation adopted in HCOS. All the selected texts belong to the genre of legal discourse and share common features of purpose, addressee, and level of formality. Therefore similar linguistic choices may be expected, also in the realm of vocabulary.

2. Binominals

The language of the law is controlled to guarantee unambiguity and compactness. Very often, however, this tendency leads to the paradox of 'verbosity for the sake of clarity', as observed by Mellinkoff (1963) and Rissanen (1999: 191-192). Set against this background, the use of binominals seems to be a natural development of a lexical strategy (to repeat in order to disambiguate). On closer inspection, however, the phenomenon of binominals is by no means so straightforward.

Binominals, or "repetitive word pairs" in Koskenniemi's early discussion (1968), are traditionally defined as "two or more words or phrases belonging to the same grammatical category, having some semantic relationship and joined by some syntactic device such as 'and' or 'or'" (Bhatia 1994: 143). Indeed, the importance of co-ordinating devices in English binominals, regardless of the historical period under investigation, had been stressed by Koskenniemi (1968: 11): "[the] connective word [...] appears to be an almost indispensable element in the repetitive phrases proper".

Binominal constructions have lately become the centre of attention in multi-word-unit analyses, which study the form and function of single cognitive entries comprising more than one lexical item (Cowie 1992; Arnold 1995; Lewis 1997; Moon 1998; Sag *et al.* 2003; Wang 2005; Abu-Ssaydeh 2005). They have also fallen within the scope of psycholinguistic investigation as *frozen expressions*, or *freezes* (Fenk-Oczlon 1989). A discussion of possible motivations behind specific vocabulary choices in binominals found in Scots and English sixteenth-century legal texts is offered by Bugaj / Włodarczyk (2006).

Binominal constructions would most often appear in answer to a stylistic need of a particular genre, e.g. in classical rhetorics (Koskenniemi 1968: 12) or legal discourse (Hiltunen 1990: 55; Bhatia 1993, 1994; Tiersma 1999: 32). Crystal observes that "[t]he words were often paired to cover distinct nuances, thereby avoiding ambiguity; but sometimes the pairing seems to be no more than a more emphatic expression of a single meaning; and sometimes it seems to be just a stylistic habit, perhaps fostered by its undoubted rhythmical appeal[4] in oral performance" (2004: 152).

Of course not every sequence of two conjoined syntactic elements of the same status and grammatical category is a binominal. Typical binominals are characterised by institutionalisation / lexicalisation, fixedness,[5] and non-compositionality,[6] (Moon 1998: 43-44). If such strict formal rules apply, the choice of lexical items cannot be haphazard, either. The factors which influence vocabulary choice include:

- the phonetic shape of the items, defined by means of alliteration (going back to Anglo-Saxon, found not only in poetry but also e.g. in Wulfstan's didactic prose (Koskenniemi 1968)); length and rhythm (not only of the phrase itself but also of the sentence); and rhyme (Greenbaum / Quirk 1990: 278; Gustaffson 1974, 1976; Wright / Hay / Bent 2005);
- semantics, traced by means of frames, synonymy and various sense relations (generalisation, narrowing, etc.) (Lambrecht 1984);
- historical reasons: for instance, the theory that after the Norman Conquest English terminology could not cover new registers, connected with feudalism and new administration (Mellinkoff 1963);[7]
- fashion, that is a preference for lexical ornamentation and expanding stylistic range with alternative items, frequently taken from different etymological sources.

---

4 On binominals serving as a mnemonic technique see Hiltunen (1990: 25-26) and Koskenniemi (1968: 116-118); on repetition in verbal rituals see Danet / Bogoch (1992: 152-154) , who discuss Anglo-Saxon curses in wills.

5 Word-order is usually fixed, although this rule can be violated: see the examples in Bugaj / Włodarczyk (2006).

6 Kiparsky (1976) discusses non-compositionality as a property of phrases whose meaning does not emerge as a sum of individual meanings.

7 This could perhaps account for the frequent presence of two etymological sources in word pairs (French and English, Latin and English), but it neglects the long history of binominal expressions in English, dating back to OE legal texts (wills, charters), translations and literature (Koskenniemi 1968).

As a result, the etymological background of potential components of a binominal pair could play an important role in their selection.

As for their semantic structure, in reduplicative phrases the following patterns occur (Wang 2005: 515):

- exact repetition (*sees* > *sees*)
- reiteration (*sees* > *sights*)
- relexicalisation (*sees* > *glimpses*)

These, however, viewed in the context of binominals, need to be narrowed down to *reiteration* and *relexicalisation*, supplemented with Koskenniemi's observation that binominals do not necessarily consist of synonymic pairs. On the contrary, one finds expressions which are antonyms or whose meanings complement each other in some way (1968: 90-94). The following sense-relations have been traced in binominals:

- near-synonymy, e.g. *clean and unblemished*
- contiguity of meaning,[8] e.g. *whole and sound*
- complementation of meaning, e.g. *towns and villages*
- antonymy, e.g. *waxing and waning*
- enumeration, e.g. *men, women and children*[9]

Such an array of formal and semantic combinations opens up a fascinating field of study, especially if one considers the possibility of lexical choices from different etymological backgrounds.

---

8 Near-synonymy is sometimes difficult to distinguish from contiguity, the basic difference being that contiguous meaning is joined by some kind of metonymy, e.g. *gave and granted, fire and flame* (Koskenniemi 1968: 91-92).

9 In fact, expressions comprising more than two items are quite frequent in my material, but fall outside the scope of this contribution.

3. Foreign influences on Scots

Vocabulary choices can be treated as a response to the stylistic needs of a given genre. Görlach (2002: 99) summarizes two motivating principles behind lexical enrichment in Scots and English:

> [F]rom the 15C onwards, developments on the levels of both Scots and English syntax and lexis represent a response to two different though related requirements:
>
> 1) to create precision, flexibility and perspicuity in complex arguments and text types newly established or taken over from Latin or French, in order to meet the demands of increasingly written communication including the formal ranges;
> 2) to add elegance and rhetorical beauty to a language insufficiently equipped to express the stylistic variation demanded by the principle of adequacy (decorum), and to approximate it to Latin – the ideal model of stylistic perfection (as it was conceived).[10]

It is worth noting that the number of borrowings into legal prose (burgh court records) decreases with time. According to HCOS studies, the period up to 1400 is the richest in loanwords and only 12.5% of loanwords are introduced in the period between 1500-1570 (Meurman-Solin 1993: 227-231). This pattern will be reflected in the etymological patterns visible in the acquired vocabulary as different sources were available at different points in time.

As for the etymological structure of the Scots lexicon in the early sixteenth century, three basic sources of vocabulary can be traced:[11]

1) a Germanic core, stemming from Northern Anglo-Saxon dialects, to which an Old Norse element can be added (see Macafee 2003: 50);

---

10 On stylistic levels and syntactic complexity in Scots, see Romaine (1981).

11 Cf. Hughes (2004: 8) for the situation in English: "The original Anglo-Saxon register typically survives in simple, direct, common terms. The word stock of the conquering Norman French elite commonly consists of a more courtly, refined register, while the Latin and Greek borrowings of the late Medieval and Renaissance periods are of a learned, abstract register."

2) Latin-influenced items, visible in education and deemed a major source for Scots up to 1400 (Görlach 2002: 11), which marks a difference between the Scots and the English of the period; Latin had an impact on the formation of the national (near-) standard and was a prolific source for literary diction;[12]
3) French-influenced items, present in Scotland for different reasons than in England, which was ruled through French for a substantial amount of time. In Scotland, French was the language of the Auld Ally, and a source of literary inspiration. However, according to Görlach, "the French contribution to Scots has not yet been systematically analysed, nor has it been compared with the French element in English" (2002: 128). This opens new ground for research, but complicates etymological enquiries into the Scots lexicon.

Interestingly, in a perspective narrowed down to Scottish religious and administrative text types, Görlach identifies a different etymological set for the period in question: Latin, Scots and Scottish English, with a primary position of Scots (see graph, 2002: 26). It would be more precise, however, to restrict the Scottish English element to the second half of the sixteenth century. In the discussion of Scots official writings, "[a] contrastive analysis with EModE texts would be needed to show how far the *distinctness* of Scots documents is [...] found in lexis and syntax" (Görlach 2002: 57, emphasis added). He goes on to observe that burgh records were much more conservative[13] than other text types in giving way to the English influence: "Scots features were apparently supported by tradition and the formulaic character of the texts" (2002: 57).

12 Interestingly, Görlach observes that after 1603 "only legal terms – which have a strikingly high percentage of items in Scottish law – exhibit a notable Latin component" (2002: 127); the same author adds, however, that seventeenth-century texts should be seen as written in Scottish English, and not in Scots.

13 As pointed out by Marina Dossena (p.c.), the arguable use of the term 'conservative' should be interpreted as "resistant to anglicising tendencies" and not as "preserving obsolete linguistic structures".

4. Binominals in early-sixteenth-century Scots administrative records

4.1. The data and research procedure

Combining the two sources: material from the HCOS and an unedited MS of burgh court records from Wigtown, has allowed me to compare the results obtained from a corpus analysis with what has emerged from the analysis of a full text. The corpus approach has the advantage of referring to a larger quantity of material by means of representative samples of particular texts, thanks to which one can study the parallels between the samples and draw generalisations. This kind of study may profit from contrasting its results with an analysis of a running text. Applying this method, my study seeks to validate the outcomes of corpus-based research.

For the corpus-based part, all official records from Period 1 in the HCOS have been chosen (ca. 60,000 words, time span 1519-1556),[14] whereas the text-based part studies an extract of running text taken from the Wigtown Burgh Court Book (1512-1534), of approximately the same length as the corpus sample (see Table 1).[15] In choosing data, I focused on phrases corresponding to a narrower version of the definition of a binominal given by Bhatia (1994: 143). As a result, only binominals built of two lexical items and coordinated by 'and' (or an ampersand '&') were selected. More complex or borderline cases (longer expressions, other possible conjunctions, etc.) were excluded from this preliminary project.

---

14 The Peebles records SREC1D were excluded as they represent the second half of the century (1555-1573), when anglicising trends could have altered the linguistic choices already (see Devitt 1989). Tracing the extent of anglicisation towards the end of the sixteenth century could possibly be the next step of inquiry into discourse-specific Middle Scots lexis.

15 The whole extant text of the Wigtown Burgh Court Book amounts to approximately 75,000 words. However, there is a gap in the MS after the first 99 folios, with occasional pages surviving from before the next continuous part (folio 238b onwards). As this last part is written in a different hand and exhibits different morphological features (see Bugaj 2004), the last folios have been excluded from the current investigation.

SPARL1: Acts of Parliament	1525-1555	30844
SREC1A: Stirling burgh records	1519-1529	11019
SREC1B: Carnwath barony book	1523-1524	1748
SREC1C: Edinburgh burgh records	1540-1551	3140
SREC1E: Aberdeen burgh records	1519-1556	10557
Total word count		57308
Wgt Burgh Court Book	1512-1532	58880

Table 1. Dates and word counts.

The selected expressions were then checked for their etymological background on the basis of the *Dictionary of the Scots Language* (DSL) online.[16] However, establishing the source of a given item was not always a simple process, especially in relation to Romance sources – Anglo-French (AF), Old French (OF), Latin (L), Medieval Latin. Some words could have been filtered through or borrowed from English. Similarly, the etymology of derivations can be difficult to establish. Bearing that in mind, the etymologies were identified as follows:

1) Gmc: this label was given to all words originally present in the Germanic core vocabulary of Scots, deriving from northern ME dialects and ON; in order to avoid confusion as to the original source of cognates, I treated early borrowings into OE as Gmc vocabulary stock, e.g. *law*, OE *lāgu*, prehistoric ON. **lagu*;
2) OF: all entries which came directly from OF or through AF, or were recorded earlier in ME than in Scots and had a French origin in English; for the words which had been present in ME, ultimately coming from Latin but present in ME and OF, I assumed OF as the etymological source;[17]

16 DSL combines the *Dictionary of the Older Scots Tongue* (+ Supplements) (DOST) with the *Scottish National Dictionary* (SND), thus offering the most comprehensive coverage of Scots lexis to this date.

17 In her detailed historical account of the Older Scots language in the DOST, Macafee (2002) divides lemmata into 5 major etymological categories: OE, F, F/L, L and Scandinavian. It is problematic, though, to make clear-cut distinctions between the early sources of Romance vocabulary in Scots or English. As pointed out by Roger Lass (p.c.), the original source of Romance vocabulary in ME (whether Latin, Old French, Medieval Latin or some other intermediate source) is largely undecidable. A suggested solution was to call

3) L: all entries which came directly from Latin, either classical or medieval; for the words which did not have a ME counterpart, or were used in EModE or later, I assumed a direct Latin background; sometimes entries in the DSL would make it clear that a certain item was typically Scots, stemming directly from Latin;

The items were then grouped according to their linear structure A+B, as reflected in the etymological components, e.g. Gmc+X, or X+Gmc. There are two possible points of focus in the phrase. Adopting the assumption that the first element in a binominal structure is less salient for various reasons, and is 'contextualised' by what follows (made more precise, explained, narrowed down, expanded – depending on the actual need),[18] I intended to see what etymological sources render most A-elements, most B-elements, and whether there was a prevailing pattern of etymological combinations.

4.2. Synonyms and other semantic relations vs etymology

The degree of transparency of the source for contemporary users is difficult to establish. It follows from the structure of binominals that the source language was important – if it had not been so, users would not have repeated the same idea twice in various 'languages'. Even if one accepts the idea of repetition as a genre-specific feature (Wang 2005), conspicuous etymological contrasts in the phrase may point to a conscious, rule-governed lexical choice. Consider the following:

---

all these items *Latinate* without further differentiation. To avoid confusion with later Latin loans, I chose to mark such items as OF in this paper.

18 See Gustaffson (1976), for the view that the first element is less marked than the second. Koskenniemi (1968: 91) observes that: "the general principle of varied repetition [...] governs the form of these phrases, and distinguished them from direct repetitions of the same word. [...] At least a slight difference in emphasis is probably always there. The second term, [...], is perhaps felt to carry more weight, even semantically"; see also the studies on the irreversibility and 'frozenness' of binominals, e.g. Malkiel (1959), Hiltunen (1990), Tiersma (1999).

➢ binomials relying on the same language source:

burghis [Gmc] *and townis* [Gmc]
rasit [Gmc] *nor vpliftit* [Gmc]
signe [OF] *& subscriptioun~* [OF]
verifeit [OF] *and prouin* [OF]
creatiounis [L] *and haill protocollis* [med. L.]
su~mond~ [pop. L.] *& p~ponit* [L. Only Sc.]

➢ binomials relying on a contrast between Romance and Germanic sources:

counsale [OF] *& help* [Gmc]
clame [OF] *&* ry^{t} [Gmc]
bundin [Gmc] *& oblist* [OF]
answerit [Gmc] *and concludit* [L]
obseruit [L] *and keipit* [Gmc]

➢ binomials comprising earlier Romance vs later Latin:

publist [OF] *and p~clamit* [L]
cassit [L] *and annullit* [OF]
ratifeit [med. L.] *and appreuit* [OF]

The overall pattern seems to favour binominals built of elements from the same linguistic background (see Section 4.3).

On the structural side, the following patterns may be observed (Wang 2005: 515; Koskenniemi 1968: 90-94):

1) exact repetition – in this case the repetition may involve a change of a grammatical structure or marking:

tratour [OF] *or tratouris* [OF]
c~mittit [L.] *or to be c~mittit* [L]
land [Gmc] *or landis* [Gmc]

2) addition of a new contextual element:

a) near-synonymy: *airis* [OF] *and successouris* [OF]
b) contiguity of meaning: *soune* [Gmc] *and air* [OF]
c) complementation of meaning: *provest* [Gmc] *and ballies* [OF]
d) antonymy: *nouther in writ* [OE] *nor word* [OE]

The semantic perspective constitutes yet another possible approach to the study of binominals. Perhaps correlations may be found between a given semantic structure and the etymological sources of its elements.

4.3. Research results

The results show that binominals have an important place in the lexical characteristics of legal discourse (around 7% of all vocabulary items is involved in their formation, see Table 2), which adds to the formulaic character of the text type.

Place	*Word-count*	*Percentage of all vocabulary*
Parliament	30844	7.2%
Stirling	11019	6%
Aberdeen	10557	7.6%
Carnwath	1748	4.5%
Edinburgh	3140	5.8%
Total	57308	6.9%
Wigtown	58880	2.4%

Table 2. Percentage of the whole word-count involved in the formation of binominals.

There are, however, important differences between texts, most notably between the Wigtown records and the corpus samples. In the Wigtown text, there is a large number of binominals (c. 1500 tokens), but this construction does not seem to be as conspicuous as in the case of the HCOS excerpts. There may be various reasons for that difference: most probably, it reflects regional discrepancies in the adoption of text-type specific features, which suggests a considerable degree of freedom in shaping administrative records in the south-west of Scotland. This would confirm Meurman-Solin's observation that in "non-literary genres such as legal documents, [...], a high degree of heterogeneity [...] ha[s] also emerged in regional and local varieties used in relatively restricted areas" (2000: 155). It is vital to notice, though, a high degree of similarity between the samples taken from all the other localities.

As for the preference in the etymological sources for the two elements of a binominal pair, the graphs in the Appendix present the tendencies observed. Clearly, with regard to types, there are two major sources which render the first element in a binominal pair: Germanic (Gmc) and Old French (OF), with OF taking the first place in a binominal in the majority of cases.

The soaring percentage of Gmc elements in the Carnwath records is due to the fact that the sample is quite short and happens to include a large number of Gmc+X expressions in terms of percentages, whereas in terms of raw counts the number is not impressive at all (7 types). In the case of Wigtown, however, the strong position of Gmc elements in binominal constructions distinguishes this text from the corpus material.

As for tokens, the pattern is not straightforward anymore. It follows from the data that there may be more types of binominals starting in an OF element, but it is often the ones with a Gmc beginning which are used most frequently (e.g. in Edinburgh and Stirling). The Carnwath records again seem to privilege the Gmc+X construction, which in fact has been caused by a frequent employment of the phrase *callit* [Gmc] *& no^{t} co~perit* [OF] (15 occurrences out of all 26 binominals found).

The Wigtown records privilege binominals starting in a Gmc element also in the number of tokens. Over 60% of all binominals begin in a word of Gmc origin, which again draws attention to the peculiarity of this court book in terms of lexical choices.

It is the second element in a binominal pair, though, which receives more focus in prosodic and, possibly, in semantic terms (Gustaffson 1976; Koskenniemi 1968: 91). The records analysed provide a very clear picture of preference for an OF second element in a binominal, especially in terms of tokens (Fig. 4-4a). As for types, there is a tendency to use more binominals finishing in an OF element. The pattern prevails in all the locations but for Stirling, where there is a slight preference for a Gmc element in the final position.

In this context, Stirling is similar to Wigtown, where again Gmc elements come to the fore in the final position within a binominal expression. It is interesting to see Latin as a source for the final element of over 20% of types in the Acts of Parliament. This practice may indicate a difference between the two subtypes of administrative records: burgh court books and parliamentary acts. The language of the latter may be more ornate and open to Latin borrowings because of the stylistic and rhetorical requirements of a legal act, as opposed to legal proceedings, whose task is mostly keeping records of meetings and reporting on the actions taken.

The most transparent picture emerges from the etymological analysis of the second element in terms of tokens. Clearly, there are more phrases with an OF element closing the structure, which is a tendency apparent even in the Wigtown records, where in general Gmc vocabulary seems to be preferred.

The preference for the OF element in all texts, be it the corpus material or the Wigtown records, proves the special value of the French lexicon in the creation of legal language. Even though there may be variation in terms of the types of fixed phrases, depending on the place, the actual instances of phrases with a final French element are most frequent everywhere.

The data can also be analysed with regard to the preference for a given combination of etymological sources for a particular grammatical category in a text. Table 3 shows that the most common combinations for all grammatical categories are OF+OF and Gmc+Gmc, the latter winning only slightly over the non-homogenous combination Gmc+OF. The clear preference in binominal expressions for both elements coming from the same etymological source contradicts the traditional view according to which binominals are an answer to comprehension demands in a multilingual setting. It seems that the point of using two closely related lexical items in a fixed expression was not to 'translate' but to enhance meaning or add a stylistic effect (see Tyrkkö, this volume).

	Nouns	*Verbs*	*Adjs*	*Advs*
Aberdeen	OF +OF	OF+OF	Gmc+OF (types) Gmc+Gmc= OF+OF (tokens)	Gmc+Gmc
Carnwath	Gmc+Gmc	Gmc+OF	–	–
Edinburgh	OF+OF Gmc+OF (tokens)	Gmc+OF = OF+Gmc (types) OF+OF (tokens)	–	–
Parliament	OF+OF	OF+OF	OF+OF= Gmc+Gmc (tokens)	Gmc+Gmc
Stirling	Gmc+Gmc (types) Gmc+OF (tokens)	OF+Gmc	Gmc+Gmc	–
Wigtown	Gmc+Gmc	Gmc+Gmc= Gmc+OF (types) Gmc+OF (tokens)	Gmc+Gmc	–

Table 3. Combination preference.

5. Concluding remarks

In the legal register, the place of binominals is quite important (about 7% of all vocabulary items is involved in their formation), which confirms the formulaic character of the text type. A variety of sense relations can be observed between the elements of the pairs (relexicalisation, clarification, contextualisation, etc.), which opens an intriguing perspective for future research. A close semantic analysis of binominal pairs seems to be the next logical step in the search for motivations for the choice of particular items within the phrase.

For early sixteenth-century Scots administrative records, there are three major sources of vocabulary: Germanic, Old French and Latin. In terms of frequency of etymological backgrounds, OF is the most prolific source. Surprisingly, however, Germanic vocabulary is also widespread in binominals, which would underline the historical continuity in legal procedures reflected in the use of native lexicon.

In terms of etymologies Wigtown does not conform to patterns of text-type creation which hold true for other contemporary burghs. It may point to the fact that text-type conventions were still fluctuating, or that administration was not centralised enough to establish a stylistic standard in the whole country. It is intriguing, though, that such distant places as Aberdeen would follow the same conventions as the Lowland burghs, whereas the scribes in the Gallovidian burgh of Wigtown would favour a markedly different choice of constructions.

My study can be seen as a starting point for a comparative investigation of discourse-specific lexis in Scots and English texts; it can also be extended to later periods to gauge the effects of anglicisation in vocabulary choices in legal Scots. Ideally, it can lead to an analysis of historically motivated discourse-specific vocabulary of the late Middle Ages and early Renaissance in Britain (influx of new lexis as a result of conquests, sociocultural background, foreign influences and native traditions), as compared to the discourse-specific lexical practices of the period in Continental Europe.

References

Primary sources

DSL = *Dictionary of the Scots Language*. Edinburgh: Scottish Language Dictionaries, www.dsl.ac.uk.

HCOS = *The Helsinki Corpus of Older Scots*. Ed. by A. Meurman-Solin, 1993. Helsinki: University of Helsinki. It includes:

SPARL1 = *Acts of the Parliaments of Scotland*, 1124-1707, vols. I-XII. Edinburgh, 1814-1875.

SREC1A = *Extracts from the Records of the Royal Burgh of Stirling*, 1519-1752. Vol. I. Ed. by R. Renwick. Glasgow 1887.

SREC1B = *Carnwath Barony Court Book*, 1523-1542. HM Register House, SRO Rh 3/10/892. Ed. by K. Williamson, 1995. Edinburgh: University of Edinburgh.

SREC1C = *Extracts from the Records of the Burgh of Edinburgh*, 1403-1589. Ed. by J.D. Marwick, 1871. Vol. ii. Edinburgh: Scottish Burgh Records Society.

SREC1E = *Extracts from the Council Register of the Burgh of Aberdeen*, 1398-1570. Vol. I. 1844, Aberdeen: Spalding Club.

Wigtown Burgh Court Book 1512-1534. Unedited and unpubl. MS, a typed transcript by Alfred Truckell. Dumfries Local Archives.

Secondary sources

Abu-Ssaydeh, Abdul-Fattah 2005. Variation in Multi-Word Units: The Absent Dimension. *Studia Anglica Posnaniensia* 41, 125-146.

Agutter, Alex 1990. Restandardisation in Middle Scots. In Adamson, Sylvia / Law, Vivien / Wright, Susan (eds) *Papers from the 5th International Conference on English Historical Linguistics.* Amsterdam: Benjamins, 1-11.

Arnold, Doug J. 1995. Multi-word Units: Collocations and Idioms. At www.essex.ac.uk/linguistics/clmt/MTbook/HTML/node55.htm.

Barisone, Ermanno / Maggioni, Maria Luisa / Tornaghi, Paola (eds) 2004. *The History of English and the Dynamics of Power.* Alessandria: dell'Orso.

Bhatia, Vijay 1993. *Analysing Genre. Language Use in Professional Settings.* London: Longman.

Bhatia, Vijay 1994. Cognitive Structuring in Legislative Provisions. In Gibbons, John (ed.), *Language and the Law.* London: Longman, 136-155.

Bugaj, Joanna 2004. *Middle Scots Inflectional System in the South-West of Scotland.* Frankfurt a/Main: Peter Lang.

Bugaj Joanna / Włodarczyk, Matylda 2006. At 'the wylle and plesur' of 'kyng and soueraine': In Search of Motivations for Binomials as Markers of Legal Discourse. In Dziubalska-Kołaczyk, Katarzyna (ed.) 2006. *IFAtuation: A Life in IFA. A Festschrift for Professor Jacek Fisiak on the Occasion of his 70th Birthday*. Poznań: Wydawnictwo Naukowe Uniwersytetu im. Adama Mickiewicza, 85-97.

Corbett, John / McClure, J. Derrick / Stuart-Smith, Jane (eds) 2003. *The Edinburgh Companion to Scots*. Edinburgh: Edinburgh University Press.

Cowie, Anthony P. 1992. Multiword Lexical Units and Communicative Language Teaching. In Arnaud, P.J.L. / Bejoint, Henri (eds) *Vocabulary and Applied Linguistics*. London: Macmillan, 1-12.

Crystal, David 2004. *The Stories of English*. New York: Overlook.

Danet, Brenda / Bogoch, Bryna 1992. "Whoever alters this may God turn his face from him on the day of judgement". Curses in Anglo-Saxon Legal Documents. *Journal of American Folklore* 105, 132-165.

Del Lungo Camiciotti, Gabriella 2004. Discursive Practices and Public Use of English at the End of the 14th Century. In Barisone *et al.* (eds), 95-103.

Devitt, Amy 1989. *Standardising Written English. Diffusion in the Case of Scotland 1520-1659*. Cambridge: Cambridge University Press.

Diller, Hans-Jürgen 2001. Genre in Linguistic and Related Discourses. In Diller, Hans-Jürgen / Görlach, Manfred (eds) *Towards a History of English as a History of Genres*. Heidelberg: Winter, 3-43.

Fenk-Oczlon, Gertraud 1989. Word Frequency and Word Order in Freezes. *Linguistics* 27, 517-556.

Goodrich, Peter 1990. *Legal Discourse: Studies in Linguistics, Rhetoric and Legal Analysis*. London: Macmillan.

Görlach, Manfred 1998. *Even More Englishes. Studies 1996-1997*. Amsterdam: Benjamins.

Görlach, Manfred 2002. *A Textual History of Scots*. Heidelberg: Winter.

Greenbaum, Sidney / Quirk, Randolph 1990. *A Student's Grammar of the English Language*. London: Longman.

Gustaffson, Marike 1974. The Phonetic Length of the Members in Present-Day English Binominals. *Neuphilologische Mitteilungen* 75/4, 663-677.

Gustaffson, Marike 1976. The Frequency and 'Frozenness' of some English Binominals. *Neuphilologische Mitteilungen* 77/4, 623-637.

Hiltunen, Risto 1990. *Chapters on Legal English. Aspects Past and Present of the Language of the Law*. Helsinki: Suomalainen Tiedakademia.

Hughes, Geoffrey 2004. Changes, Registers and Institutions. In Barisone, Ermanno *et al.* (eds), 5-23.

Jones, Charles (ed.) 1997. *The Edinburgh History of the Scots Language*. Edinburgh: Edinburgh University Press.

Kiparsky, Paul 1976. Oral Poetry: Some Linguistic and Typological Considerations. In Stolz, Benjamin A. / Shannon, R.S. (eds) *Oral Literature and the Formula*. Ann Arbor: University of Michigan, 73-106.

Koskenniemi, Inna 1968. *Repetitive Word-pairs in Old and Early Middle English Prose*. Turku: Turun Yliopisto.

Lambrecht, Knud 1984. Formulaicity, Frame Semantics, and Pragmatics in German Binominal Expressions. *Language* 60/4, 753-796.

Lewis, Michael 1993. *The Lexical Approach*. Hove: Language Teaching Publications.

Macafee, Caroline 2002. A History of Scots to 1700. In Dareau, Margaret *et al.* (eds) *A Dictionary of the Older Scots Tongue*. vol. 12. Oxford: Oxford University Press, xxxi-clii.

Macafee, Caroline 2003. Studying Scots Vocabulary. In Corbett *et al.* (eds), 50-71.

Malkiel, Yaakov 1959. Studies in Irreversible Binomials. *Lingua* 8, 113-160.

McClure, J. Derrick 1994. English in Scotland. In Burchfield, Robert (ed.) *The Cambridge History of the English Language*, vol. 5. Cambridge: Cambridge University Press, 23-93.

Mellinkoff, David 1963. *The Language of the Law*. Boston: Little, Brown and Co.

Meurman-Solin, Anneli 1993. *Variation and Change in Early Scottish Prose*. Helsinki: Suomalainen Tiedeakatemia.

Meurman-Solin, Anneli 2000. Change from above or from below? Mapping the *Loci* of Linguistic Change in the History of Scottish English. In Wright, Laura (ed.) *The Development of Standard English 1300-1800*. Cambridge: Cambridge Univ. Press, 155-170.

Meurman-Solin, Anneli 2001. Structured Text Corpora in the Study of Language Variation and Change. *Literary and Linguistic Computing* 16/1, 5-27.

Meurman-Solin, Anneli 2003. Corpus-based Study of Older Scots Grammar and Lexis. In Corbett *et al.* (eds), 170-196.

Meurman-Solin, Anneli 2004. Data and Methods in Scottish Historical Linguistics. In Barisone *et al.* (eds), 25-42.

Moon, Rosamund 1998. *Fixed Expressions and Idioms in English: A Corpus-based Approach*. London: Clarendon.

Rissanen, Matti 1999. Language of Law and the Development of Standard English. In Taavitsainen, Irma *et al.* (eds). *Writing in Nonstandard English*. Amsterdam: Benjamins, 189-203.

Romaine, Suzanne 1981. Relativisation, Syntactic Complexity and Stylistic Levels in Middle Scots. *Folia Linguistica Historica* 2, 56-77.

Sag, Ivan A. *et al.* 2003. Multiword Expressions: A Pain in the Neck for NLP. At <lingo.stanford.edu/pubs/WP-2001-03.pdf>.

Tiersma, Peter M. 1999. *Legal Language*. Chicago: Univ. of Chicago Press.

Wang, Shih-ping 2005. Corpus-based Approaches and Discourse Analysis in Relation to Reduplication and Repetition. *Journal of Pragmatics* 37, 505-540.

Włodarczyk, Matylda. 2005. Pragmatic Aspects of Reported Speech: Early Modern English Court Records and Trial Transcripts. Unpublished PhD thesis. Adam Mickiewicz University, Poznań.

Wright, Saundra K. / Hay, Jennifer / Bent, Tessa 2005. Ladies first? Phonology, Frequency and the Naming Conspiracy. *Linguistics* 43/3, 531-561.

Appendix

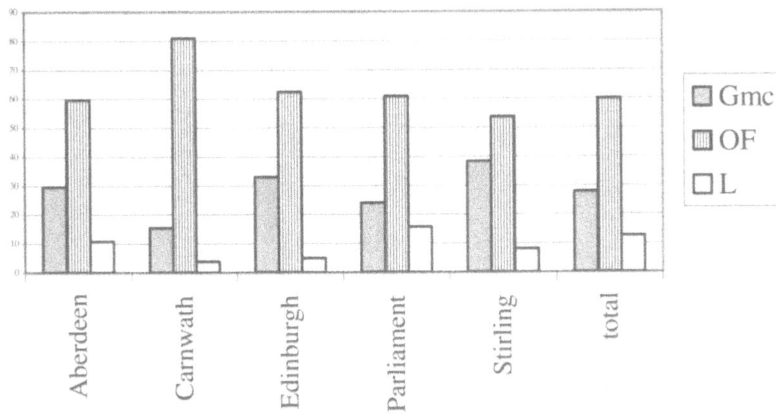

Figure 1. The etymological background of the first element in a binominal (types).

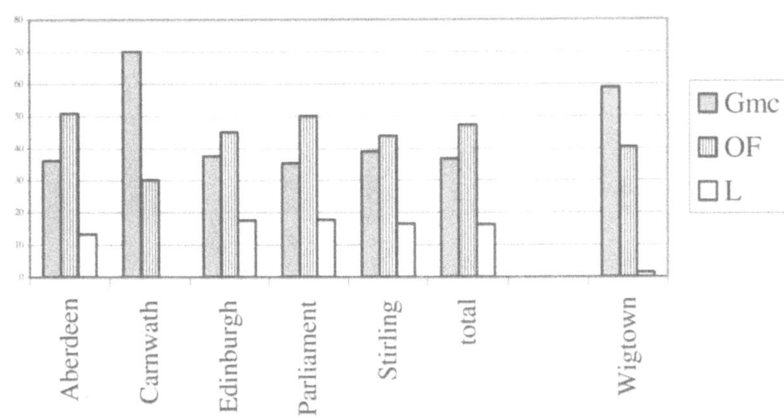

Figure 1a. The etymological background of the first element in a binominal, Wigtown included (types).

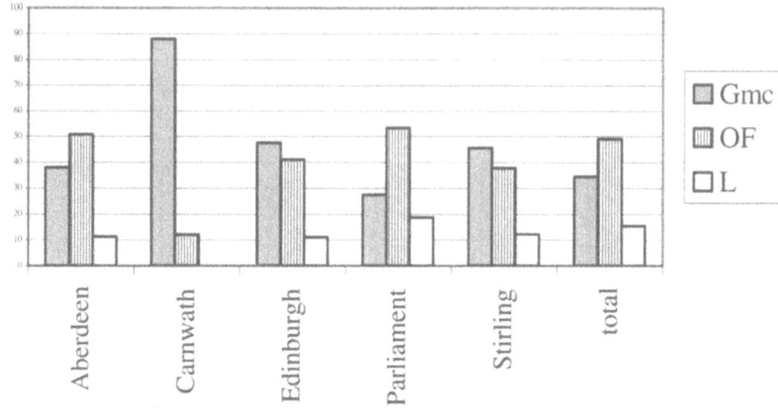

Figure 2. The etymological background of the first element in a binominal (tokens).

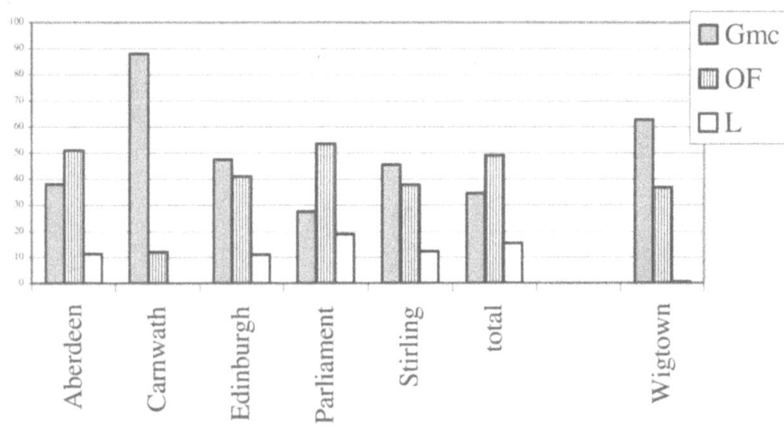

Figure 2a. The etymological background of the first element in a binominal, Wigtown included (tokens).

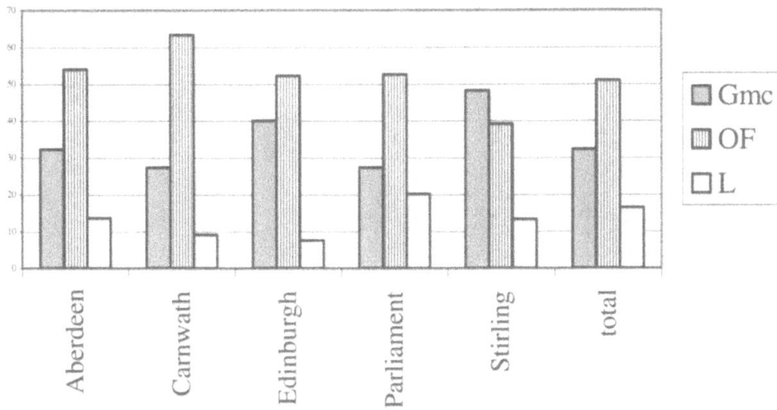

Figure 3. The etymological background of the second element in a binominal (types).

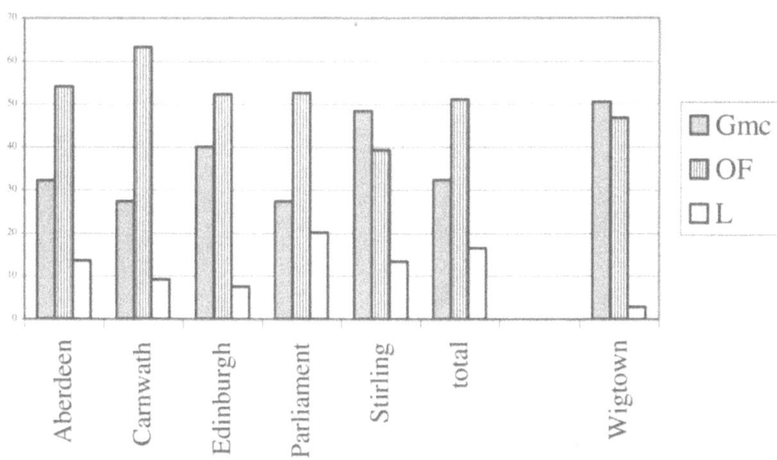

Figure 3a. The etymological background of the second element in a binominal, Wigtown included (types).

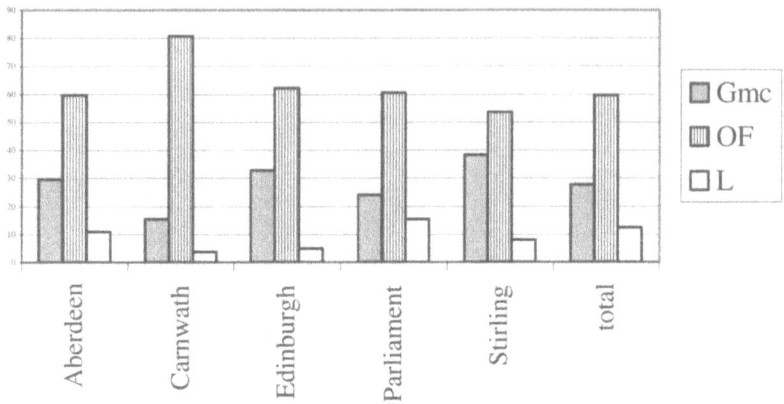

Figure 4. The etymological background of the second element in a binominal (tokens).

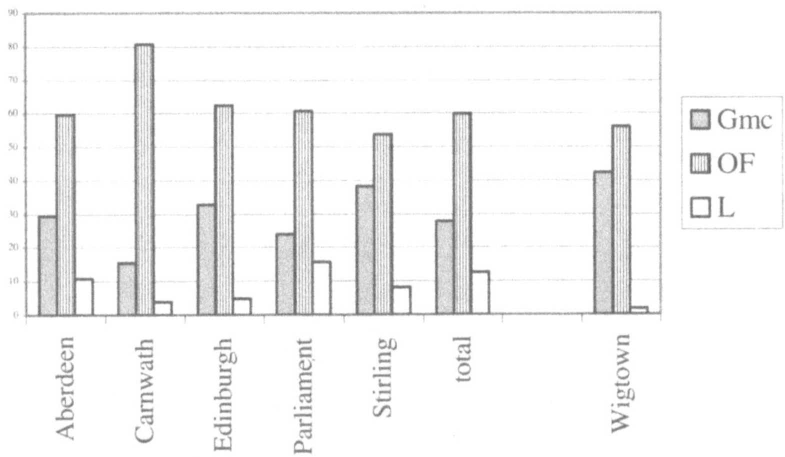

Figure 4a. The etymological background of the second element in a binominal, Wigtown included (tokens).

Marjorie Lorvik

Language Contact and Lexical Variation in the Context of the North Sea Timber Trade

1. Introduction

Imagine a scenario where traders of different nationalities and language backgrounds can each use his own language, and indeed his own language for a specific purpose (LSP), yet still understand and be understood, at least to a certain extent. Could this have been the situation in the North Sea area, for example in the sixteenth and seventeenth centuries? In this study I look at the possibility that mutually intelligible terminology cognates were used in the North Sea timber trade during this period. An examination of a small sample of terms relating to the timber trade found in the cargo inventory of the *White Lamb* of Silt from 1698 and in customs records for Scotland and Norway from this period would seem to indicate that each participant in this trade might have been able to use his own language in order to do business with traders from the other North Sea countries.

However, trade in several other goods in the North Sea area was also fairly intensive and had a long history. Undoubtedly, traders from the different countries in this area would have been strongly motivated to understand each other in order to do business. This kind of long-term, international trade situation could typically have laid the foundation for a trade pidgin to develop, particularly if the languages in question were not related to each other in any way and were mutually unintelligible. Yet no such North Sea trade pidgin has been found, which would seem to indicate that the degree of intelligibility was such that trade could be conducted quite adequately without. Furthermore, the fact that there could have been mutually intelligible LSPs might be an indication that there was also some degree of mutual intelligibility in the languages for general purposes (LGPs) in this area at the time, based both on languages that were cognate and on the communication strategies employed. After all, trade also has its social aspects, which are normally conducted in LGP.

1.1. Research background

While researching possible lexical transfer from Norwegian to Scots, in particular to the dialect of the North-East of Scotland known as the Doric (see Lorvik 2005), I came across a small number of timber terms in a cargo inventory and some customs records written during the time of the so-called Scottish Trade between western Norway and Scotland from approximately 1550 to about 1700. The references to these timber terms were found in economic history material in both Norwegian and English that I had consulted for information about trade contact between the two countries during the relevant period. The cargo inventory, recorded in connection with the sale of the *White Lamb* of Silt on 8 October 1698 to Alexander Pyper at Burntisland in Scotland was found in Smout (1999: 57).[1] He states that, though the boat was owned in Denmark and sold by its master "Mr Mathias Jowgarsone", the cargo "can only be from Norway". The inventory reads as follows:

> Item 21 dusson of daills of 12 foot of length.
> 20 dusson of daills of 10 foot of length.
> 3 dusson of wrack daills at 12 foot length.
> 2 dusson of double12 elns trees,
> 45 ten eln trees.
> 2 ½ dusson of double oak garrons [posts or beams]
> 4 dusson of sparrs
> 4 fathom of burnwood.
> 3 half barrels of tarr.
> A dusson of oak knee-heads.
> 4 dusson of schoolles [shovels].
> 3 great trouches.
> 2 small trouches and
> 2 long oars.

1 The source for the inventory given by Smout (1999: 57) is: SRO (Scottish Record Office): Misc. Proc. Bundle 228, Alexander Pyper MSS, papers relating to the sale of the *White Lamb* of Zilt.

This set me off on an investigation into why so many of these Scots terms seemed to be so similar or fairly similar to their counterparts in the Norwegian material. The results turned out to be, at the very least, intriguing.

1.2. Linguistic background

First of all, it should be pointed out what is meant by the term Scots. Modern Scots, one of Scotland's three indigenous languages alongside Gaelic and English, is descended, like modern English, from Anglo-Saxon, or OE. However, while Standard English is descended from the Mercian dialect of OE, Scots derives from the Northumbrian dialect, which in OE times was in use from the River Humber to the River Forth (McClure 1995: 6). The Viking invasions of the north and east of England, among other places, which began just before 800, resulted in considerable permeation of the Northumbrian dialect by Norse, especially since Anglo-Saxon and Norse were both descendants of Germanic and thus "probably largely mutually comprehensible in the way that divergent dialects of the same language are today" (Milroy 1997: 318). By 1000 Scotland had two main languages: Gaelic over most of the country and Norse-influenced Anglo-Saxon in the south-east, which was the fore-runner of Modern Scots. Then, of course, there was also Norwegian in Orkney, Shetland and Caithness (Murison 1977: 2). However, what was of greater significance for the strong Scandinavian element in Scots even today was the period of immigration from Northern England into Scotland during the Normanisation process in the eleventh, twelfth and thirteenth centuries. Scots can therefore be said to be a secondary contact dialect with Norse.

The development of Norwegian had been somewhat complicated by the fact that Danish had become the official written language in Norway in 1536. This was a result of the Dano-Norwegian Union which was established in 1380 and lasted till 1814. Danish was also the variety used by the new cultural elite and the wealthy commercial middle class, who became the most powerful class in the seventeenth and eighteenth centuries (Almenningen 2002: 47). In addition, the introduction of the printing press in 1643 in

Norway pushed the written Norwegian language even closer to Danish, parallel to the developments in the spoken language of the Norwegian upper classes.

However, as far as the Hanseatic period is concerned, according to Jahr (1999: 128), we should not think in terms of discrete languages in Scandinavia. Instead, there would have been a range of spoken variants of Low German, which would have enabled merchants, almost the only people to travel at the time, to do business in the local variety (Jahr 1999: 128). This was probably also the language situation at the beginning of the Early Modern period. Norwegian and Danish dialects, with their large proportion of Low German lexis, would both have been spoken in Norway at this time. Indeed, even today Norwegian and Danish, not to mention Swedish, are largely inter-intelligible as part of the North Germanic dialect continuum (Wiggen 1995: 65).

2. The etymological investigation

The etymological investigation of the timber terms that follows here looks at the relevant items listed in the inventory of the *White Lamb* from 1698. These were obviously standard items of cargo, as Lythe's (1976: 148) discussion of some sixteenth century terminology from the trade in timber between Scotland and Norway includes many of the same terms as are in the inventory.[2] The aim is to uncover evidence of any possible linguistic influence or possible borrowing specifically from Norwegian. In some cases other North Sea languages have also been mentioned, but in this limited examination a synchronic analysis of the etymology of each word has not been possible. Consequently, I have used the same approach described by Macafee (1997: 292) concerning the lexis of Older Scots: "Where an Old Norse etymon is not known, a word may nevertheless be identified as Scandinavian by the discovery of parallels in the modern Scandinavian languages and dialects".

2 For an investigation of the terms from Lythe see Lorvik (2003).

The various dictionaries that have been consulted are referred to by the following abbreviations:

CSD	*Concise Scots Dictionary*
DSL	*Dictionary of the Scots Language* (online)
EO	*Etymologisk ordbok (over det norske og danske sprog)*
ESO	*Engelsk stor ordbok* *(Norwegian-English, English-Norwegian)*
JSD	*Jamieson's Scottish Dictionary*
NEO	*Nynorsk etymologisk ordbok*
OED	*Oxford English Dictionary* (online version)
SND	*(Compact) Scottish National Dictionary*
SNO	*Store norske ordbok*

First of all, the terms were examined in *OED*, *SND*, *DSL*, *CSD* and *JSD*. In most cases either a form currently used in English, e.g. *deals* for *deallis*, or the etymological background for the Scots form could be identified. Norwegian dictionaries were also used, to check whether the same or a similar form of the term is still used and to try to find the etymology of the word in Norwegian as well as possible cognates in other North Sea languages. *OED* has proved an excellent source of information about cognates in the other languages in question, i.e. Dutch, (Low) German, Danish and Swedish. However, there have also been references to these languages in some of the entries in the Norwegian dictionaries. In other words, through examining the etymology I hoped to find whether any of these terms indicated that the terminology used in the North Sea timber trade might have been fairly international and mutually intelligible.

It is, however, important to remember that the situation for Scots generally is very similar to that described by Melchers (1987: 314) regarding the Low German and Dutch element in the Shetland dialect. Her conclusion is that "it is often impossible to determine whether a word is of Norse, Scots or [Dutch / Low German] origin. It is also difficult to tell whether an alleged [Dutch / Low German] word is a direct loan or whether it has reached the dialect via Scots or Norn". Borrowings into Scots have very often left little evidence concerning their route of entry. This means that direct contact between Scots and Dutch could be the explanation, or contact between Dutch and Norwegian first, and then contact between Norwegian and Scots. In a business context we can imagine that, although each participant used his own cognate variant of the term in question, they could still have been understood sufficiently for a deal to be reached.

3. Terms from the inventory

The terms from the inventory of the *White Lamb* are listed below in the order in which they appear in Smout (1999: 57). Four of these terms also occur in Lythe (1976: 148), though perhaps in a slightly different form. These are marked with an asterisk and Lythe's comment on the meaning is included beside the term.

**daills* – "boards […] an obvious word"

CSD lists *dale* and *dell*, meaning "deal, a plank", dated late sixteenth – early seventeenth century and located in Angus and Perth. *OED* lists the third meaning of *deal* as "a slice sawn from a log of timber", and gives one of the various alternative spellings as *dail* in Scots. It explains that in the timber trade in Britain and North America a *deal* is understood to have specific dimensions for it to be thus classified. What are more interesting, however, are the cognates that are listed: modern Dutch *deel*, modern German *diele.* Also mentioned is the fact that the etymology of *deal* can be traced back to c 1400, when its root was introduced from Low German. Smout states that *deals*, "sawn fir planks of varying lengths", were the type of timber most frequently listed in the customs books in the second half of the seventeenth century (1963: 155). Lythe points out that this is one timber term that would be familiar to a modern timber merchant (1976: 148).

No similar form has been identified in the Norwegian etymological dictionaries with the same meaning. However, Lillehammer, in his study of Scottish and Dutch trade with the Agder and Rogaland coasts of Norway, in the south and the south-west of the country, mentions that cut boards were called *delar* (2001: 20). Also in Næss (1920: 24) the word *deler* occurs, and is explained as meaning *bord*,

Norwegian for "boards, planks". In *Den norske sjøfarts historie*, reference is made first in English to "Norway deales" and then in Norwegian to "*Norgesdeler*" being delivered to Yarmouth (Worm-Müller 1923/I: 367).

The headwords here have therefore most probably been one of several very similar forms found previously in the various languages of the North Sea trade area for many centuries. The term has apparently not survived in Norwegian, perhaps as it was only used in the commercial context of the timber trade.

wrack daills

Wrack, as a modifier of *daills*, refers to the quality of the deals, that is, the fact that they were damaged or unusable for their primary purpose as boards or planks. This is the meaning given in *OED* for the adjective *wrack*, and an example dated 1584 is quoted, where *wrak* is used to refer to fish that is sorted according to quality. The etymology of the word is traced back to Middle Low German *wrak* and the same form is given for Middle Dutch. *CSD* lists the adjective *wrack*, dated late sixteenth century, as meaning "damaged".

SNO gives the Norwegian form as *vrak* with the same meaning of "damaged, unusable", which indicates that the combination *wrack daills* would probably have been intelligible for Norwegian traders. No doubt this also applied to traders from Holland, Germany and Sweden as well.

dussons

Obviously this plural noun has to do with the quantity of the item in question, rather than the item itself, but it would still have been an important word in the world of the timber trade. *OED* gives OF *dozeine* and *doseine* but there is no mention of OE or ON cognates. However, there was most probably a form of this word used in the other North Sea languages centuries ago, judging by its existence in them today. The *Online English to Dutch to English Dictionary* lists *dozijn*, *EO* lists Norwegian *dusin* and Swedish *dussin* as derived from Middle High German *Dutzend*. It also mentions French *douzaine* and the Latin word for twelve, *duodecim*. We may thus assume that this word would have been mutually intelligible among traders of various linguistic backgrounds around the North Sea and even further afield. *CSD* gives a variety of alternative spellings, including *dussone*, the earliest date given being the late fifteenth century.

eln

Also this word does not refer to a type of timber as such, but, as a quantifier, it had a very important function in the trade. An ell, according to *OED*, was "a measurement of length representing the distance from the elbow or from the shoulder to the wrist or to the finger-tips". Indeed, as *OED* points out, *elbow* itself is a compound containing the word, which in Old Teutonic meant arm or fore-arm. *OED* gives cognates from Dutch, German, Swedish and Danish and gives examples showing just how far back its etymology can be traced, including OE and Old High German. *DSL* lists several spelling variants, including *elln(e), ellin* and *ellen.* Smout comments in connection with the ship's inventory that "an 'eln' or 'ell' was about a yard" (1999: 57). *CSD* comments that the Scots ell was approximately four-fifths of the English ell and dates this late fifteenth century.

In the article in Norwegian by Næss (1920), which deals with the Norwegian trade in timber with Scotland from Sunnhordland in western Norway, he quotes from the customs books from the first half of the seventeenth century. One specific example he gives of the lengths involved contains the plural form *alner* (36), meaning 'ells', which must have been quite comprehensible for the Scottish traders, and indeed for those from other North Sea countries.

**garronis* – "[...] vague"

The first meaning given for *garron*, according to *CSD* is "wooden beam". It is dated sixteenth century onwards and is now found in Orkney and Caithness. *SND* defines it as "a length of squared timber, a beam". *OED* gives Old Scots *garro(u)n,* "short wooden beam", from 1543, but this is of uncertain origin, perhaps Old North French *garron*, from Old French *jarron*, branch of a tree. *JSD* lists *garron nails*, "large nails of different sizes", but there is no explanation for *garron*. *CSD* also gives *garron nail* dated sixteenth century, and defines it as a large nail or spike used to fix the body of a cart to its axle. *The Shetland Dictionary* (Graham 1999) lists *garron* as meaning "a large, square-shaped nail". There would seem to be no cognate in Norwegian.

**sparris* – "spars ('an obvious word')"

OED defines *spar* as "1. one of the common rafters of a roof, [...] 2. a pole or piece of timber of some length and moderate thickness". Cognates include Dutch and West Frisian *spar*, Norwegian, Danish and Swedish *sparre*, indicating a broad geographical area of use. It is also suggested there may be a connection with French (modern French dialect *épare*). *CSD* also lists *spar* meaning quite simply a spar, dated sixteenth century. *EO* lists Norwegian *sparre*, and mentions Dutch *spar* meaning "beam". The entry also says that French *espars* is derived from Germanic. Norwegian *sperre*, meaning "bar, barrier; rafter" is also mentioned. These are obviously all cognates.

**burnewood* – "emphasises the role of wood as fuel in this period"

Burnwood occurs in one example in *OED* from 1898 under *knapple*, in a list of types of timber exported from Norway. It is listed in *CSD*, explained quite simply as wood for burning, and dated sixteenth-nineteenth century. It also occurs in *JSD*, meaning "wood for fuel". Smout comments that "if there was a little space left after the loading [of a ship], it was often filled with a fathom or two of 'burnwood', perhaps used for kindling in Scottish households" (1963: 156). His use of quotation marks would suggest that this was not a standard term, to his knowledge, other than in the customs records. Smout makes another reference to the word (1999: 43) when, in commenting on an entry in the so-called import books of the early 1680s for Prestonpans, he points out that "'Stowage' was firewood, or 'burnwood'". *OED* defines *stowage* as "that with which a vessel is or is to be stowed".

There is no apparent corresponding compound in Norwegian in the dictionaries listed above. However, both elements in the Scots compound noun have cognates in Norwegian: *brenne*, "burn" and *ved* "wood". The word is in fact listed in the *Norsk ordbok: ordbok over det norske folkemålet og det nynorske skriftmålet* (1966). It has also been confirmed by native speakers that *brenneved*, referring to wood for heating purposes, exists certainly in dialect, differentiating it from *settved*, which is listed in *SNO*. Here the latter is defined as wood which was used in mines to heat up the rock face so that the stone would break into pieces. The headword here would therefore have been easily understood by both Norwegians and Scots, and may very possibly be a loan translation into Scots.

It is also possible that there is a cognate in Dutch, for example, as "burn" translates as *branden* and "wood" as *woud*. However, judging by the entry for "wood" in *Cassell's English-Dutch Dutch-English Dictionary,* by far the more frequent form used in Dutch is *hout*, which is a cognate of German *holz*.

fathom

This word refers to the measurement of quantity of the *burnwood* in the inventory. Although nowadays it indicates a unit for measuring water depth, the *OED* lists the third meaning as "a measure of length" and the fifth meaning as "a certain quantity of wood", with examples from the late sixteenth century. It also gives some compounds, including *fathom-wood* with the cognates Swedish *famnved* and German *fadenholz*. Smout uses *fathom* as a unit of measurement in his comments on *burnwood* as stowage (see above).

DSL lists *fadom* with several other variants, from ME and eME *fadom(e),* and gives the second meaning as "a measure of peat, wood, etc." The earliest example given is *faddomes* dated 1564 from Orkney and Shetland. *CSD* dates *faddom* from the late sixteenth century and *fadom* from the fourteenth-nineteenth century. *JSD* lists *fadom* and gives the Icelandic *fadm-r*. Here the meaning is given in Latin as *quantum mensura se possunt extendere lacerti cum manibus.*

EO gives the Norwegian word *favn* and Swedish *famn* and states the English cognate is *fathom* and the Dutch *vadem*. In *SNO* *favn(e)ved* is given as meaning "wood 60 cm long". The headword and all its cognate variants would have been mutually intelligible.

tarr

Again this is not directly a timber term, but it is closely related, as it is used for coating and preserving timber. It is therefore hardly surprising it is part of the cargo of the *White Lamb*. According to *OED* the OE form was *teru*, Middle German *ter* and *tere,* and ON *tara*. The Danish form is *tjære*, which is also the modern Norwegian form.

According to *DST tar, tarre* and *ter* are some of the forms this word has had. An example dated 1661 has the same spelling as the headword here. *CSD* dates the form *tar* as sixteenth century.

schoolles

As Smout indicates that this word meant "shovels", tracing its origin was much easier than it might have been. As a matter of fact, *OED* gives *s(c)hoole* as one of the various dialect forms of *shovel* and includes an example from a letter written by Robert Burns in 1785, "Ye'd better taen up spades and shools". *JSD* also lists it quite simply as meaning "shovel". *CSD* lists *shuil*, dated seventeenth century, as an alternative for *shovel.*

SNO gives the Norwegian form *skuffel*, from Middle Low German *schuffele. EO* lists *skovl* and mentions Swedish *skyffel* and Dutch *schoffel* as cognates.

trouches

DSL gives *trouch* as one of the several Scottish forms of English *trough. OED* confirms that it is a Scots dialect form, the OSc form being *trog.* Middle Dutch is given as *troch(-gh)*, Dutch *trog,* Middle High German *troc (trog),* ON *trog,* Norwegian dialect *trog, trugh* (*traug,trau*), Swedish *tråg*, and Danish *trug*. The Indo-European form **druko-* is also mentioned, given as a derivation of "*dru,* TREE, wood, timber; primary meaning 'wooden vessel'". Under *troch* in *CSD,* a variety of forms are given, including that of the headword here, the earliest dating from the late sixteenth century. *SNO* lists Norwegian *trau* as meaning *trough*, from ON *trog.*

It is most probably this type of object that is being referred to in the ship's inventory, the same meaning of modern *trough,* which *OED* defines as a long narrow open box-like vessel, often used to contain water. Some other areas of use listed include washing, brewing and tanning. However, it is unclear whether there is any direct connection with the other timber products in the cargo. In any case, this word appears to have been in use, in one of its many variants, for a very long time, and would most probably have been intelligible to timber traders from different North Sea countries.

4. The Scottish Trade: historical background and linguistic traits

The so-called Scottish Trade in timber between Scotland and Norway, in particular the Stavanger area, which lasted approximately from 1550 to1700, came about as a result of a plentiful supply of timber and improved technology in the shape of the water-driven sawmill on the east side of the North Sea and a demand for timber for shipbuilding and other constructions on the west side. The evidence of this trade, recorded in customs books on both sides of the North Sea in Scots and Norwegian, shows clearly its scope and intensity. A list from 1567 with details of the vessels that had visited Ryfylke is the oldest source of information in Norway about this trade. Of a total of 38 vessels, there were at least 28 that were Scottish (Lillehammer 1986: 101). Some Scottish ships called frequently in the main trading period, from March till September, and stayed for several days, in some cases even weeks. One case in 1642 stands out in particular, when David Walker stayed for four months while repairs to his ship were carried out (Lillehammer 1986: 107). Lillehammer comments that "Frequent calls and long stays [...] must have resulted in close connections between buyer and seller, between the east and the west side of the North Sea" (1986: 107).

However, towards the end of the seventeenth century, the Scottish Trade was reduced considerably. Trade in timber then increased in other areas of Norway and it also became triangular in nature, between Scotland, Norway and Holland (Lillehammer 1986: 108), with goods such as salt, herring, butter, textiles, pottery and silver being exported from Holland (Lillehammer 2001: 8). Dutch traders buying timber were by far in the majority in certain parts of the south-west coast of Norway but they certainly faced strong competition from Scottish traders. In addition, though fewer in number, there were also trading vessels from Denmark, Germany, England, France and Spain (Lillehammer 2001: 9-10).

Heerma van Voss (1996: 14) describes more generally how trade in the North Sea stretches as far back as the eleventh century, when there was regular contact for this purpose between Flanders and

England. Then the Hanseatic League became established by the mid-fourteenth century, "uniting nearly two hundred cities in the Baltic and North Sea areas for the protection of their trade" (Heerma van Voss 1996: 14). That there was close cultural contact is borne out by the extensive correspondence from this period that has survived, from private as well as public figures, which demonstrates that "travellers along the North Sea and Baltic coasts understood each other very well using each their variant of *Nederduitsch*" (Heerma van Voss 1996: 15).

The existence of a North Sea cultural area in the early modern period, resulting from intensive trade, is also discussed by Heerma van Voss (1996: 22), who comments on trade contacts as follows (1996: 24):

> Norway provided wood for shipbuilding and other construction work in England and Holland. [...] Denmark exported grain and cattle to Holland and the North German cities. Denmark and Norway exported fish, as the Dutch did to the Baltic coasts. The fish trade carried the salt trade in its wake, which in its turn was followed by the trade in Mediterranean products. The Scottish lowlands and the Shetlands depended for many primary products upon imports from England, Holland and Scandinavia. In exchange the Shetlands exported fish to Bremen and Hamburg. Lowland Scotland exported grain to Norway and Rotterdam, coal – as did Northern England – and salt to the whole of Northern Europe and coarse textiles, fish and cattle.

Heerma van Voss points out that the Orkneys and Shetlands formed a natural part of this area, and so did the Baltic shores. He states that "in every Baltic port, Scots, English and Dutch were to be found" (1996: 29). With regard to wealth in the area, however, he draws a distinction between "the urbanised, modern and rich southern part (England, Flanders, the Dutch Republic and the German North Sea coast) and the less developed and poorer northern part (Scotland, Denmark, Norway)" (1996: 29). Yet wealth apart, the various countries all belonged to the same business, cultural and, in a loose sense, linguistic area. There was also trade contact with the rest of the world but contacts "within the North Sea area were much more frequent and intensive than those with the outside" (1996: 24).

4.1. A North Sea trade pidgin?

Considering the fact that the North Sea countries were carrying on trade for a long period of time, the question arises whether there is any evidence of the existence of a trade pidgin in use in this area. However, to date there certainly seems to be no trace of any having existed.

As we know, a pidgin is a language that emerges as a result of contact between people who have different mother tongues and who need to communicate. It usually consists of a mixture of the languages in question (Broch / Jahr 1981: 12). A trade pidgin or trade jargon is usually a rudimentary pidgin, such as existed in Finnmark in the north of Norway as a result of the Pomor Trade between Russians and Norwegians. This trade pidgin was known as *russe-norsk,* and was in use from the late eighteenth century until the First World War, a period of roughly 140 years. From the written material available, it consisted of about 390 words, which is typical of a trade pidgin (Broch 1992: 133). Being able to speak Russian or Norwegian did not mean traders could then understand *russe-norsk.* They had to learn it, as demonstrated by a conversation between a Russian captain and a Norwegian fisherman, transcribed by a Norwegian customs officer in Tromsø in 1921 (Broch 1992: 130).

The Pomor Trade and the Scottish Trade have in common the fact that they lasted for approximately the same period of time, that is, about 140 years. Yet no equivalent of *russe-norsk* has been found relating to the Scottish Trade, or in fact relating to any of the trade activity in the North Sea area. This could suggest there was actually no need for one. Each trader could probably use his own variety of North Sea language, and understand and be understood sufficiently for business to be conducted satisfactorily. Of course, this would no doubt have required considerable communication skills from each participant.

4.2. Mutual intelligibility of terms and communication in the North Sea timber trade area

In the absence of a special trade pidgin in the North Sea area, we should then consider whether traders were able to communicate sufficiently, each using their own language, in order to do business. This situation would probably have involved a specialised terminology for the product in question and some commercial language no doubt for the administrative side of things, that is to say, LSPs to a greater or lesser degree. However, at the pre-industrial time under discussion, the products traded would in many cases not have been of a very complex technical nature and so certainly with regard to fairly straightforward products such as grain, salt and timber, the transaction would most probably not have required a great number of technical terms to be used.

The sample above of eleven timber terms in Scots is indeed a small one, but it may well indicate the linguistic situation at a time when there was a hive of trade activity in the North Sea area, particularly in timber. Ten of the eleven terms discussed above can be said to be cognate with terms used in Norwegian and other North Sea languages (the exception being *garronis*). They could then very possibly have been mutually intelligible.

Braunmüller (1997: 365-366) details four different ways of communicating when two languages are in contact with each other. The fourth one is of greatest relevance for our scenario:

> In some cases, it might be sufficient just to understand what the others say without trying to achieve an active command of the respective language. When this process continues, there is a fair chance of becoming a passive bilingual. This strategy works, however, only when the languages involved are genetically rather closely related and when all participants in language contact see an obvious advantage in keeping their own language and in learning correspondency rules as well as some lexical variants, at least for the moment. Later on, they may decide to become active bilinguals, to continue in this way, or just to take over some loans which have proved to facilitate interlingual communication.

Braunmüller comments that speakers "have to concentrate on the most meaningful elements in an utterance, e.g., nouns and verbs" (1997:

369). He also states that "trading contacts differ from other forms of linguistic contacts by being single-minded: the only thing that really counts is to sell as much as possible and to establish firm contacts for expanding these trading relations in the future" (1997: 367). He compares the current situation in Scandinavia with language use at the time of Hanseatic influence in these terms:

> each participant uses his/her own language, vernacular or even dialect when meeting another Scandinavian. He/she can be sure of being fairly well understood, not only because of some more or less perspicuous linguistic coincidences, but also on the basis of feeling themselves to be members of a common Scandinavian community. These historically and pragmatically motivated dimensions in inter-Scandinavian contact situations of the present-day correspond with the economic aims and vital interests of all trading participants in the Baltic area and in western Norway at the time of the Hansa (in the Middle Ages and later). (Braunmüller 1997: 367)

Jahr (1999: 127-128) points out that the Scandinavian languages and Low German between 1350 and 1500 were mutually intelligible, with Low German occupying a prestigious position. No common 'Hansa language' had developed, and there has been no evidence found so far of a pidgin or mixed language of Low German and the Scandinavian languages during the Hanseatic period (Jahr 1999: 134). There did exist early Russian textbooks for Hanseatic merchants, Low German and Russian obviously being two discrete languages, but no equivalent textbooks for the Scandinavian languages have ever been found (Jahr 1999: 130). Nor have any such textbooks been discovered for Scots or English. This suggests that they probably did not exist, most likely because there was no need for them.

Trade in the early modern period between Norway and Holland, which was another member of the Low German language community, was also extensive. In that connection Sogner (1996: 192) states:

> Fundamental for all contact is language. But there are several levels of linguistic understanding, and communication may be possible although the command of a language is imperfect. The period in question – the early modern period – is still fundamentally oral. People managed their everyday affairs quite well without being able to write. In such a culture, people in general had a better ear and a higher sensitivity of hearing than we [sic], and their comprehension of foreign languages may even have exceeded our own.

It must be remembered that the languages under discussion on the eastern coasts of the North Sea were more closely related to each other than Scots was. Nevertheless, Scots was still reasonably closely related through its Germanic roots in Old Northumbrian, and no doubt bore a far greater resemblance to them in and immediately after the Middle Ages than is the case today. The traders would therefore have had no need of a pidgin, as they may have been what Braunmüller (1997: 366) calls "passive bilinguals". Communication would have been based on their understanding what the other party was saying, without their having to speak the other party's language. Perhaps Scottish traders might have been in an even better position than those from much further south in England when it came to understanding the other North Sea languages. Heerma van Voss says that "in general the dialects of English spoken along the North Sea coasts, especially north of the Wash, have more in common with other North Sea languages than standard English" (1996: 27). Scots would have been in the same category.

Another point to consider is that during the sixteenth and seventeenth centuries in Europe local variants of a language within the same state were not necessarily mutually intelligible. At that time in Europe nation states and their standard languages were still emerging, and political and linguistic borders were not always as clearly drawn or as stable as today. Consequently, a greater degree of willingness and ability would have been required in order for people to understand a language variety not quite the same as, or even rather different from, their own.[3] Braunmüller (1997: 372) comments that the Hanseatic traders in particular would have encountered various languages and dialects at the different market places and fairs they visited. As a result, they would have been linguistically flexible and able to adapt to new linguistic situations, gradually becoming what he calls "passive multilinguals" (1997: 372). Very probably this was also the case in the hectic sixteenth- and seventeenth-century North Sea trade.

Gass / Selinker (1994: 206) describe a study of non-native speaker (NNS) speech which showed that native speakers (NSs)

---

3 Indeed, Norway today is a very good example of this, as Norwegians are accustomed to communicating with other Norwegians who may speak quite a different, not very intelligible, dialect altogether.

understood more of NNSs' speech, the more they were exposed to it. There were particularly three factors that helped listeners understand: (1) familiarity with a particular NNS, (2) familiarity with non-native speech in general, and (3) familiarity with the discourse topic. This would no doubt have been the discourse scenario in the North Sea timber trade.

5. Conclusion

This is indeed a very limited etymological investigation focusing on only eleven items related to the North Sea timber trade in one ship's inventory dating from 1698. Yet it is striking to see that as many as ten of these eleven terms were most probably mutually intelligible, certainly among traders using North Sea languages such as Scots, Norwegian, Dutch, Danish and Low German. Judging from the absence of a trade pidgin similar to *russe-norsk* we may then perhaps conclude that traders in fact managed to do business through semi-communication and passive bilingualism.

Apart from the one term, *burnewood,* any other trace of lexical borrowing among the terms investigated, particularly from Norwegian into Scots, is really non-existent, certainly in the period in focus here, from the mid sixteenth century to approximately 1700. Instead, there seems to be some evidence that a form of mutual intelligibility, certainly as far as trade is concerned, may have existed in the North Sea area. In order to draw such a conclusion with greater certainty, however, much more extensive research is obviously required.

References

Almenningen, Olaf *et al.* (eds) ^{6}2002. *Språk og samfunn gjennom tusen år: ei norsk språkhistorie / A Millennium of Language and Society: A History of the Norwegian Language.* Oslo: Universitetsforlaget.

Braunmüller, Kurt 1997. Communication Strategies in the Area of the Hanseatic League: The Approach by Semi-communication. *Multilingua* 16/4, 365-373.

Broch, Ingvild / Jahr, Ernst Håkon 1984. *Russe-norsk – et pidginspråk i Norge (Russo-Norwegian – A Pidgin in Norway).* Oslo: Novus.

Broch, Ingvild 1992. Russe-norsk – Pomorhandelens eget språk / Russo-Norwegian – The Language of the Pomor Trade. In Niemi, Einar (ed.) *Pomor: Nord-Norge og Nord-Russland gjennom tusen år / Pomor: North Norway and Northern Russia over a Thousand Years.* Oslo: Gyldendal Norsk, 129-138.

Edmonston, Thomas 1866. *Etymological Glossary of the Shetland and Orkney Dialect.* London: Asher & Co.

Gass, Susan M. / Selinker, Larry 1994. *Second Language Acquisition.* Hillsdale, NJ: Erlbaum.

Graham, John 1999. *The Shetland Dictionary.* Lerwick: The Shetland Times.

Heerma van Voss, Lex / Roding, Juliette (eds) 1996. *The North Sea and Culture.* Hilversum: Verloren.

Jahr, Ernst Håkon 1999. Sociolinguistics in Historical Language Contact: The Scandinavian Languages and Low German during the Hanseatic Period. In Jahr, Ernst Håkon (ed.) *Language Change: Advances in Historical Linguistics.* Berlin: Mouton, 121-139.

Lillehammer, Arnvid 1986. The Scottish-Norwegian Timber Trade in the Stavanger Area in the Sixteenth and the Seventeenth Centuries. In Smout, Thomas Christopher (ed.) *Scotland and Europe 1200-1850.* Edinburgh: John Donald, 97-111.

Lillehammer, Arnvid 2001. Skottar og hollendarar på Agder- og Rogalandskysten. In *Kontakten mellom Agder og Holland på 1600- og 1700-tallet.* Flekkefjord Museum, 4-24.

Lorvik, Marjorie 2003. Mutual Intelligibility of Timber Trade Terminology in the North Sea Countries during the Time of the 'Scottish Trade'. *Nordic Journal of English Studies* 2/2, 223-243.

Lorvik, Marjorie 2005. The Distinctiveness of the Doric: Home-grown or Imported? *Scottish Language* 24, 31-63.

Lythe, Samuel G.E. (1960) 1976. *The Economy of Scotland in its European Setting 1550-1625*. Westport: Greenwood Press.

Macafee, Caroline I. 1997. Older Scots Lexis. In Jones, Charles (ed.) *The Edinburgh History of the Scots Language*. Edinburgh: Edinburgh University Press, 182-212.

McClure, J. Derrick 1995. *Scots and its Literature*. Amsterdam: Benjamins.

Melchers, Gunnel 1987. On the Low German and Dutch Element in Shetland Dialect. In Ureland, P. Sture (ed.) *Sprachkontakt in der Hanse*. Tübingen: Niemeyer, 295-316.

Milroy, James 1997. Internal vs External Motivations for Linguistic Change. *Multilingua* 16/4, 311-323.

Murison, David 1971. The Dutch Element in the Vocabulary of Scots. In Aitken, A.J. / McIntosh, Angus / Pálsson, H. (eds) *Edinburgh Studies in English and Scots*. London: Longman, 159-175.

Murison, David 1977. *The Guid Scots Tongue*. Edinburgh: The Mercat Press.

Næss, Anders 1920. Skottehandelen på Sunnhordland. *Sunnhordland tidsskrift* VII, 23-55.

Norsk ordbok: ordbok over det norske folkemålet og det nynorske skriftmålet 1966. Oslo: Det norske samlaget.

Sogner, Sølvi. 1996. Popular Contacts between Norway and the Netherlands in the Early Modern Period. In Heerma van Voss, Lex / Roding, Juliette (eds), 185-198.

Smout, Thomas Christopher (ed.) 1963. *Scottish Trade on the Eve of the Union: 1660-1707*. Edinburgh: Oliver & Boyd.

Smout, Thomas Christopher 1999. The Norwegian Timber Trade before 1707, from the Scottish Perspective. In Lillehammer, Arnvid / Langhelle, Svein Ivar / Smout, T.C. / Bruin, Jaap (eds), *Timber and Trade: Articles on the Timber Export from the Ryfylke Area to Scotland and Holland in the 16th and 17th Centuries*. Aksdal: fagrapport / Lokalhistorisk stiftelse 1, 37-57.

Wiggen, Geirr 1995. Norway in the 1990s: A Sociolinguistic Profile. *International Journal of the Sociology of Language* 115, 47-83.

Worm-Müller, Jac S. (ed.) 1923. *Den norske sjøfarts historie*, vol. 1. Kristiania: Steenske.

GABRIELLA DEL LUNGO CAMICIOTTI

From *Your obedient humble servants* to *Yours faithfully*: The Negotiation of Professional Roles in Commercial Correspondence of the Second Half of the Nineteenth Century

1. Introduction

Specialised discourses are normally studied from a synchronic perspective in that the results of the research can be applied to the teaching of English for specific purposes. Recently, however, historical studies of English have increasingly been devoted to the investigation of communication in the law courts, in government, and in the church, with a view to tracing the history of the discursive practices of such institutions.[1]

A relatively neglected area is the history of business discourse. One reason for this may be that the merchant class is a constantly evolving community of traders rather than an institution, built on oral rather than written traditions. Business has only recently become an academic discipline and is still a rather diffuse and interdisciplinary area of studies, as opposed to a clearly defined science such as economics. With the enormous expansion of the economic and business foundations of the modern world, there is an increasing need to shed light on the history and practice of business writing (Douglas / Hildebrandt 1985). To help narrow this gap, I have constructed and investigated a small specialised corpus of correspondence from the second half of the nineteenth century, comprising commercial letters

1 Among the many contributions on discourses from a historical perspective see Hiltunen / Skaffari (2003) and Skaffari *et al.* (2005).

collected by William Anderson (1873), Teofilo Cann (1883), and Romeo Candelari (1899) for didactic purposes.

Business writing is a modern concept, yet commercial transactions have been conducted both orally and through the written word since the Middle Ages. Indeed, by the fifteenth century letter writing had become part of the daily life of the gentry as well as of the professional and merchant classes (Richardson 1985).[2] The nineteenth century is crucial for the development of written commercial communication because of rapid changes in the conditions of business and social life. In both Britain and America, a marked rise in maritime commerce led to a considerable increase in letter writing, as well as in the production and consumption of guides for native and foreign novices in the business world.

The main purpose of my research is to identify the textual construction of the international trader's social role. It aims to answer the following questions: what norms regulated the epistolary behaviour of nineteenth-century traders? Is it possible to identify linguistic strategies that typify the trader's identity as a member of the business community? And more generally, how were epistolary business relations negotiated in the late nineteenth century?

The status of the interactants is today seen as a decisive element in the distinction between professional and business discourse. While professional discourse is institutional in character, in that it also involves lay persons, business discourse exclusively concerns talk and writing between individuals who come together with the purpose of doing business (Bargiela-Chiappini / Nickerson 1999: 2). This suggests that role negotiation between trader and client takes place on equal terms. In addition, there is a corporate dimension to the identities of modern interactants. Given these premises, the question is whether the nineteenth-century businessman enjoyed a similar status and whether relations between firms were then regulated by the same norms. The research thus intends to shed light on past communicative practices in the business community in relation to changes in traders'

2 Recent overviews of letter writing from a historical perspective are presented in Nevalainen / Tanskanen (2004) and Dossena / Fitzmaurice (2006). See also the investigations conducted by Nevalainen / Raumolin-Brunberg (1996), Nevala (2002, 2004) and Tieken-Boon van Ostade (2004).

social status and business ethos. It aims to help define the modern business community by examining its historical roots in the crucial second half of the nineteenth century.

Since key elements of business communication at any time are information/persuasion and successful interaction, I have chosen to focus on interactional strategies (Del Lungo Camiciotti 2005, 2006). More specifically, in this chapter I will pay special attention to the use of modal verbs to express politeness and to their pragmatic meanings. The historicity of politeness, as a social norm and practice, is here posited. Speech communities possess linguistic means of enhancing the ego's social standing and of avoiding conflict, thereby maintaining the perceived fabric of interpersonal relationships between participants in a verbal interaction. Norms of politeness include ritualised forms of verbal behaviour, such as terms of address and appropriate ways of displaying deference towards the addressee and of carrying out certain types of speech acts, such as opening and closing an interaction. One of the most fruitful aspects of recent historical linguistic research into politeness in epistolary discourse has been the investigation of modal verbs expressing attitude towards content and interlocutors, as modality can be manipulated pragmatically in response to the needs and requirements of a given situation (Fitzmaurice 2002, 2003).

2. Materials and method

2.1. Data

As already mentioned, the research presented here aims to identify the textual construction of the social role of nineteenth-century international traders, as disclosed by linguistic and discursive strategies. I have therefore constructed a small specialised corpus comprising three letter writing guides from the latter half of the century (270 letters). Specifically compiled for Italian learners, these contain business letters of various length dealing with the routine activities of international firms (Del Lungo Camiciotti 2002).

Two manuals (Anderson and Candelari) contain original letters adapted for didactic purposes; the third handbook, instead, contains letters composed by the author. From the close reading of the paratext (title pages, prefaces and notes) we infer that all are practical books and contain little, if any, instruction to their prospective readers, who are supposed to learn by imitation the commercial letter writing style. The prefaces give us also some information as to the users of the textbooks: according to the authors, their chief aim is to provide a useful instrument for young men entering the business world or the counting house, but also teachers not versed in Commercial English might profit from their work. The overall word count is distributed as shown in Table 1 below:

Source	*No. of letters*	*Word count*
Anderson (1873)	151	67,426
Cann (1883)	73	14,577
Candelari (1899)	46	6,073
Total	270	88,076

Table 1. Number of letters and word counts per source.

The research is both quantitative and qualitative in character. First, it presents data relating to the relative frequency of particular features. The aim is to assess what features are most common and thus contribute to the conventionality of politeness and the negotiation of identities. Secondly, by examining full texts, it offers a broader perspective on the construction and interpretation of epistolary meaning and on interpersonality as a feature of the process of sending and receiving messages in the business community of the time. Since business interaction, like any type of social interaction, is historically constituted and situated in a specific cultural context, what counts as appropriate behaviour in business relations depends on those features of the interaction that are socio-culturally marked as appropriate by the historical speech community. The conscious choice of strategies is thus constrained by socio-cultural norms which change over time.

The data analysed here allow us to investigate the extent to which modes of communication become conventionalised within a particular historical tradition. The process of acquiring and using the skill of social intercourse is necessarily a practical and tacit one. To be able to master this skill appropriately, one has to interact in a social and practical context. And this was no less true in the past. How then do we get information about past modes of interaction? Instructions on good manners give us general information as to changes in social interaction. In the nineteenth century, numerous manuals of etiquette attempted to meet novel social needs as many of the newly rich sought to achieve social status worthy of their wealth. Information regarding verbal interaction in the nineteenth-century business community may be gleaned from the numerous letter-writing guides of the period. At this time the rise in literacy resulted in a notable increase in the practice of letter-writing (Barton / Hall 1999). Although large groups remained illiterate or only marginally literate, correspondence became of central importance in conducting social affairs and in establishing and maintaining business relations. This led to a significant increase in the production and consumption of business letter-writing manuals for native and foreign speakers of English.

A second point is that, since correspondence was then the most important means of long-distance communication and the only way of maintaining business relations in international commerce, the analysis of letters seems a particularly significant way of investigating modes of interaction in the community of international traders. In addition, historical model letters are especially useful for the exploration of conventionality in business interaction because commercial letters have always been produced by copying existing ones. For this reason they are highly conventionalised and the linguistic choices adopted in the correspondence under scrutiny may help delineate the identity of international traders in the Victorian period.

2.2. Methodological approach

A discourse-analytical and pragmatic approach sheds useful light on business letter-writing as an historically situated activity by uncovering rhetorical and linguistic devices used to implement the politeness strategies best suited to ways of doing business in specific periods. In order to develop a comparison across different linguistic and cultural contexts, a detailed micro-analysis of specific linguistic devices is needed. This is based on the assumption that politeness strategies are crucial in negotiating business relations at any time, even though their realisation may vary in accordance with changing conceptions of business ethos and values. In this chapter, I draw on corpus linguistics for discourse analysis and politeness theory in order to show how the choice of modal meanings contributes to the construction of the trader's identity and his interactional strategies.

Within this framework, the following procedure was adopted. I examined the electronic corpus consisting of business letters collected by Anderson, Candelari and Cann, to gain a general sense of the use of interaction markers. I then proceeded to select a specific analytical focus, the verbal expression of politeness. I chose to focus on this aspect because politeness is a key dimension of successful epistolary interaction in any period. Indeed, perceptions of politeness seem to arise in any culture or group with respect to the interpersonal relations and social conventions agreed upon among the group members at the time in which the interaction takes place (Watts / Ide / Ehlich 1992). Politeness can thus be considered an index of the overall style of approved interpersonal relation negotiation.

Recent linguistic research on politeness (Fitzmaurice 2002) has investigated specific areas of linguistic structure expressing speaker's attitude, including modal verbs considered as units with an interpersonal function. This approach looks to Biber *et al.*'s corpus-based grammatical description, which has focused attention on the grammaticalisation of stance. Stance subsumes both epistemic meaning (degree of certainty, doubt) and attitudinal meaning, the speaker's perspective on what is said with respect to the addressee and the text itself (1999: 972). Among the various linguistic resources used by the speaker in expressing stance are modal verbs, considered one of the most frequently occurring ways of grammaticalising stance in orally based genres.

2.3. Conceptions of politeness

Since conceptions of politeness are linked to its verbal expression, it may be useful to give a brief overview of the two basic approaches which have been identified in the literature (Watts 1992). From a traditional point of view the main function of politeness is essentially to coordinate social interaction and to regulate the mechanical exchange of roles and activities, by showing respect and deference towards social status. Here politeness is regulated by social norms. More recent conceptions of politeness, on the other hand, are comparatively individualistic, identifying it with various modes of avoiding or smoothing out conflict in a continuous attempt to balance face-saving and face-maintaining. Brown / Levinson (1987) posit that all interaction is characterized by concern over the other person's autonomous needs and his/her desire to be liked, as manifested in message behaviour that addresses those needs by taking into account three basic situational factors: relationship distance, relationship power, and the magnitude of the request.

As a matter of fact, both views are useful, in that they may be related to different ways of looking at social interaction. Indeed, as my analysis shows, nineteenth-century epistolary interaction, though certainly based on the traditional deferential paradigm, also offers indications that politeness strategies were shifting towards the modern way of perceiving polite business interaction.

The concept of politeness has certainly undergone considerable evolution over time. The term *politeness* in English can be traced back to the eighteenth century, when it was associated with metropolitan aristocracy and opposed to rural life and cultural provinciality. It meant a high degree of mental cultivation and elegant refinement, polished manners and classical good taste. Nineteenth-century novels suggest that politeness continued to be seen primarily in terms of class and power – the master/servant metaphor. However, it was also increasingly perceived as a mask, or means of achieving individualistic goals (Sell 1992).

2.4. The nineteenth-century business community

To gain a better understanding of the changes in business writing practice and of how the Victorian trading community conceived styles of polite interaction, it is necessary to sketch, albeit very briefly, the socio-historical background. The contemporary ideology of free trade was in no small part responsible for the increase in overseas trading activities. Indeed, by the second half of the century, British commercial policy was firmly committed to free trade, providing a stimulus to a new class of industrialists interested in international trade (Parsons 1999).

From a social angle, the Victorian business community was characterized by what has been labelled "gentlemanly capitalism" (Price 1999). Financial and rentier interests, rather than the new industrial entrepreneurs, dominated Victorian society, and industrial dynasties often aspired to landed lifestyles. In America too the nineteenth-century businessman had to acquire the patina of a gentleman, and there was always an assumption that the gentleman should possess certain refinements and niceties both in behaviour and expression (Douglas 1985: 127). Owing to the social improvement of the trading class, businessmen adopted a style of behaviour shaped by a gentlemanly concept of politeness involving deference and respect to others in the pursuit of one's aims.

3. Linguistic expression of politeness

Of course politeness is a complex combination of various means which are only partly linguistic, but there is no doubt that the collective norms of polite behaviour are reflected in linguistic choices. A traditional way of investigating respect and deference is to examine terms of address and greetings (Taavitsainen / Jucker 2003). However, a simple examination of the ways in which these are deployed in our guides sheds little light on the matter. First of all, they are highly conventionalised, indeed almost formulaic; and secondly, they reflect a conservative and formal attitude[3] rooted in the discourse of humility characterising the patron/client relation in the eighteenth and nineteenth centuries. Recipients are addressed as *Sir* (428

3 Bijkerk (2004) has shown that the closing formula *Your most humble servant*, which is common in the correspondence under scrutiny here, had already been superseded by *Yours sincerely* in eighteenth-century private correspondence.

occurrences), *Gentlemen* (351), *Mr* (339), *Mess.rs* (286), *Esq.* (182), while senders define themselves as *faithful* or *humble servants.* The closing formulae almost invariably contain adjectives related to the discourse of humility, the most common being *obedient* (192), *humble* (172) and *faithful* (14), which mostly modify *servant* (153) or *servants* (52). Adverbs such as *truly* (127), *faithfully* (27), *obediently* (15) are also widely used to reinforce the perlocutionary effect of assertions such as *I remain, very truly, Sir, your most obedient humble servant* (Anderson, letter VIII).

Terms of address used in salutations and closing formulae are thus very deferential, indicating the relative status of the interactants as utterly unequal. This usage conforms to the traditional paradigm of politeness expressed in terms of the master/servant metaphor. More interesting insights are yielded by the use of modal auxiliaries, in so far as these reveal the facework involved in establishing business relations. In other words, their usage may uncover less traditional and more symmetrical ways of interacting.

As Table 2 shows, the frequency of modals in the correspondence is quite high. This is not surprising, since letters are considered a form of written conversation and, as Biber *et al.* (1999: 979-980) explain, modals are more frequent in conversation, in that they mostly convey stance-type meanings. We may also note that the writers are reasonably uniform in their use of modals, seeing that differences between the various letter collections are only slight.

Anderson		*Cann*		*Candelari*		*Overall*	
Will	7.39	Will	1.20	Will	0.65	Will	9.25
Shall	3.13	May	0.49	Shall	0.23	Shall	3.86
May	2.30	Shall	0.48	Can	0.10	May	2.87
Should	1.36	Should	0.22	Should	0.07	Should	1.66
Can	0.91	Would	0.21	Would	0.06	Can	1.22
Would	0.90	Can	0.20	May	0.06	Would	1.19
Must	0.69	Could	0.14	Could	0.05	Must	0.88
Could	0.38	Must	0.13	Must	0.05	Could	0.59
Might	0.21	Might	0.06	Might	0.01	Might	0.29
Ought	0.07	Ought	0.03	Ought	–	Ought	0.11

Table 2. Frequency of modal verbs (normalized figures per 1,000 words).

Table 3 compares the frequency of modals in present-day English (from Biber *et al.* 1999: 486) and in the corpus under scrutiny. The decreasing order of frequency reveals interesting differences. In present-day English the modals *will*, *would* and *can* are extremely common, while *shall* is relatively rare. In our correspondence, on the other hand, *shall* is very common, while *could* is relatively rare.

However, frequency counts do not explain how the verbs were used contextually. If we want to investigate to what extent their usage mirrors contemporary traders' values and beliefs as to social intercourse, we must examine the modals' pragmatic meanings, indicated in Table 4, in order to ascertain whether they confirm the asymmetrical relationship expressed by the discourse of deference and respect, or whether they point to a more symmetrical inter-subjective way of looking at business relations.

Present day	*Nineteenth-century correspondence*
Will	Will
Would	Shall
Can	May
Could	Should
May	Can
Should	Would
Must	Must
Might	Could
Shall	Might

Table 3. Order of frequency of modal verbs in present-day usage and 19th-century letters.

Will, shall	Prediction / volition
Would	Volition
Shall, should, must, ought	Necessity / obligation
May, can, could, might	Ability / possibility
May	Permission
Would, should	Tentativeness

Table 4. Semantic/pragmatic value of modal verbs in 19th-century correspondence.

Each modal can of course express more than one meaning. However, it is possible to identify prevailing ones. *Will* and *shall* are used to express prediction and volition, depending upon their collocation with person. Prediction is signaled by *will* with 3p subjects and *shall* with 1p and 2p subjects. Volition is indicated by *will* with 1p and 2p subjects, and *shall* with 3p subjects. *Will* may also express obligation with 2p subjects, in expressions such as *You will please to receive* or *You will be so good as. Should* and *would* grammaticalise tentativeness in expressions such as *Should that be the case, It would be desirable, You would oblige us very much,* and in parenthetical comments: *I should think, We should imagine. Should* can also express direct obligation with 3p: *He should charge no commission,* while *would* can express volition with 3p subjects: *I'm certain they would do everything. Should* and *would* can also function as past tense forms of *shall* and *will*, though less often.

The frequent occurrence of *shall* and *will* is due to the fact that they are widely used in polite expressions together with their past forms *should* and (less often) *would.* They express tentative obligation and necessity particularly in combination with 1p subjects in expressions such as *We should be very happy, I should be glad, I should think you would have no difficulty, Should you wish me to, I would advise you, I would suggest, etc.*

When used with 1p subjects, *must* expresses the speaker's personal commitment or some sort of honourable obligation to do something, in expressions such as *I must request you, I must earnestly recommend.* With 2p and 3p subjects, it indicates obligation/necessity: *The molasses must be thick, You must take care. Ought* is quite rare and seems to have the same meaning as *should.*

May is much more common than *might.* It mostly expresses possibility, but also some form of tentative possibility/ability for politeness reasons, in expressions such as *I may say that, For any information that you may require, You may rest assured that.* Less frequently, *may* requests permission, as in *I write to ask whether I may so far trespass. Might* expresses a very low degree of likelihood: *Perhaps we might have been induced.*

Can and *could* express ability/possibility present and past as in *I will do the best I can for your interest, I will see what can be done, I am truly sorry I could not do better.*

According to Biber *et al.* (1999: 485), modals can be grouped into three main categories, according to their main meaning. If we compare present-day meanings with those expressed in our correspondence, as in Table 5, interesting differences emerge, which combine with differences in frequency to give us a more precise picture of the use of the verbs.

Present day	*Nineteenth-century correspondence*
Permission/possibility/ability: Can, could, may, might	Volition/prediction: Will, shall, would
Obligation/necessity: Must, should, ought	Obligation/necessity: Shall, should, must, ought
Volition/prediction: Will, would, shall	Permission: may Possibility/ability: May, can, could

Table 5. Modal meanings compared.

The high frequency of volition/prediction modals in nineteenth-century correspondence can be attributed to the text genre, as letters tend to bridge the time-gap in anticipating recipients' response or future action. However, the choice of modals is also dictated by politeness, reflecting the writers' conception of courteous intercourse and a business ethos based on formality and civility. *Would* and *should*, as well as other modals, are indeed widely used to express tentativeness, so as to hedge the force of assertions and are thus routinely employed to stress politeness. For instance, in (1) below, after the long preamble which explains the circumstances motivating the request to be paid in advance, the actual request is hedged by the polite expression *should you be disposed.* The point to be made here is the relatively high frequency of obligation/necessity modals, while those expressing possibility/ability are relatively uncommon.

(1) I regret that I cannot offer you credit, but the fact is that I have had such heavy losses of late in bad debts, that I have recently been obliged to lay down the rule to supply no more goods except for cash payments. My means are too limited, and the profits I realise too small to admit of my waiting for the settlement of accounts as I have hitherto done.
Should you be disposed to send me the money, I will see that your order is executed in a manner to give entire satisfaction. (Cann, Letter 49)

In most letters, facts tend to be asserted in a straightforward way rather than presented as possible – see (2) below, while the interpersonal bond between writer and recipient is constructed through the careful choice of polite wording and the use of obligation modals. For instance, in (3), a recommendation letter, the job request contained in the second paragraph is prudently modalised.

(2) We are now enabled to announce to you the conclusion of our sales of your consignments per ship 'Rubicon' and 'Sir Joseph Banks' and to transmit to you herewith the account sales thereof, nett proceeds altogether amounting to 917£. 18s. 2d. from which the remittances already made to you being deduced, the balance still remaining in your favour amount to 361£. 18s. 2d. as shewn by the accompanying accounts current. (Anderson, Letter CCCVIII)

(3) Our friend, M...., wishing to establish in your town, where he reckons to promote in the best way his interests becoming the partner of a rich relative of his, just come from [...] we take the liberty to introduce him to your respectable firm, as it is probable that he open a house in the [...] line in your town.
Should he not soon come into these business arrangements with his relative, he might want an employ for some time. In that case you would oblige us very much by admitting him into your house, resting assured that you would intrust your interests to an honest, intelligent and learned young man.
(Candelari, Letter 38)

The analysis allows us to conclude that modal verbs are an extensively used means of grammaticalising writers' attitude, and that they combine with other stance constructions in elaborate ways to express polite attitude in epistolary discourse.[4] In their role as attitude markers, they interact with other politeness strategies to situate the recipients of messages in such a way as to ensure that the face wants of both interactants receive appropriate attention. In this regard, we may also note that politeness tends to correlate with the letter sub-genre.[5] Modal auxiliaries thus provide an important pragmatic resource for the expression of politeness in the context of the nineteenth-century 'gentlemanly' business ethos, based on comity and courteous social intercourse. In addition, a close reading of the letters reveals that writers avail themselves of polite strategies in both

4 On the interaction of modals with other politeness strategies in the same correspondence see Del Lungo Camiciotti (2006).

5 In this corpus, two different sub-genres of commercial letters, transactions and circulars, have been identified according to the choice of linguistic and rhetorical devices (Del Lungo Camiciotti 2005).

conventional and less traditional ways. More precisely, two different means of expressing attitude towards interlocutors are deployed: firstly, traditional, formal terms of address and epistolary formulae are adopted to express respect and deference for the recipient's status; secondly, modal verbs, and notably obligation modals, express a more symmetrical, though no less polite, negotiation of trade relations.

To conclude this section, a typical exchange contained in the manual collected by Anderson (Letter CCLIX) may illustrate the style typical of the correspondence. The emphatic politeness associated with commercial correspondence is here immediately apparent in the writer's apology for the delay in shipping wines. In offering his own and his associates' further services, the writer expresses his stance in an elaborate way, in which modals are a major index of his aims. He first offers the renewal of his services and then suggests the line of action to be followed should the recipient decide to take up the offer. He is confident that his client will be well served, yet the scenario presented is one in which he must await his client's decision. The markedly tentative style aims to alleviate the impression of imposing an offer of renewed service on his client following the latter's complaint. While the first paragraph is written in a straightforward style, the following are modalised; here modal verbs, together with other linguistic and rhetorical devices,[6] provide a pragmatic resource for the expression of the politeness needed in order to secure the client's order.

> Messrs. Crawford and Dunn, Plymouth. London, 11th August, 18-.
>
> Gentlemen,
> I am this morning favored with your much esteemed letter of the 9th instant, and am extremely sorry you should consider there has been any improper delay in shipping your wines; but I feel confident that my Oporto friends acted for the best. Shipping is very scarce there, and I am informed that freights had risen to 3 guineas per ton.
> With respect to sending out the vessel you have at your port of about 350 tons: should you determine on so doing, and will consign her to my friends there, Messrs. Dawson, Forbes and Co. I am certain they would do every thing in their power to serve you, not only by giving her a preference of what wines

6 On the interaction of metadiscourse and stance markers and complex sentence structure in this correspondence see Del Lungo Camiciotti (2005).

they might have to ship themselves, but by inducing their friends to do the same.
I would suggest that, as soon as you come to a determination on the subject, you should write to my friends not to engage freight for your wines in any other vessel, and the same to the company. It is possible they may have already succeeded in obtaining freight for them, for it being known that freights were high and vessels scarce, some may have been sent out from different ports. This is a point of your consideration. Expecting your determination and orders for insurance,
I remain, truly, Gentlemen, your obedient humble servant,

WM. THOMAS

5. Concluding remarks

To conclude, let us return to the questions that motivated the present research. As to the discursive construction of the nineteenth-century trader's identity, we may note that in their choice of modals expressing obligation, commitment and volition/intention writers seem primarily motivated by politeness, as a way of negotiating interpersonal status in response to the particular requirements of the situation at hand. Their wish to reach business goals is expressed in a very polite way, much more polite than would be the case today, when directness and explicitness of goals are perfectly acceptable to the business community. A much more deferential and tentative epistolary style is deployed in these letters and indicated as model behaviour to learners of business writing. Moreover, the writers have a twofold attitude towards their addressees. On the one hand, they wish to conform to Victorian patterns of socially acceptable behaviour, governed by respect and deference towards social status; on the other, evidence of a more symmetrical relationship in negotiating the respective roles of trader and client emerges in the use of modality to mask their individualistic business goals.

As to whether politeness strategies in nineteenth-century business letters differ from those of today and of how such differences relate to changes in the trader's social status and identity, it may be said that in that period businessmen wrote in a 'gentlemanly' manner and did not use, as the contemporary business-man so often does, an impersonal style. On the contrary, they aimed to establish a relationship based on personal responsibility and mutual obligation. Indeed, the social and practical context of business communication was then completely different. Far from being a giant corporation run by a number of executives, the company was very much a family affair, with a small staff of office workers, possibly just the entrepreneur himself and a clerk or office boy. As businesses were centred around the authority of individuals, relations were based on a code of commercial gentility not dissimilar from that regulating general social intercourse. This could involve prolixity and circumlocution in writing, particularly in greetings and salutations, even though directness and simplicity of language was not avoided in the wording of the actual transactions.

Commercial letters seem to have evolved to meet the social and professional demands of the community of practice. At the same time, these needs determined the way in which letters were textually shaped and pragmatically realised. The conditions of business changed so quickly, and so radically, that a new pattern of expressing interpersonal relations emerged in the late nineteenth century. The conventional expression of humility, formality and ceremonious elaboration, typical of the master/servant paradigm, gave way to a more egalitarian relationship. Negotiation thus became more confrontational, hence in need of politeness strategies to avoid possible conflict. In short, the choice of linguistic forms, though in accordance with the dictates of time and fashion, reveal the attempt, on the part of traders, to enhance their standing with respect to others.

The findings of this study are based on the analysis of a small specialised corpus (270 letters) and thus shed light on the epistolary practice of the business community in the second half of the nineteenth century. The data may further be compared with modern modes of communication in order to assess changes in modes of negotiation as well as cross-cultural differences and continuities over time. Again, the data's contribution to the definition of business writing as a specific practice may be enhanced by comparison with original letters from the same period, such as those being collected for the Corpus of Nineteenth-Century Correspondence (cf. Dossena 2004, 2006, and Dury 2006).

While the benefits of synchronic research in business discourse are obvious, it is hoped that the diachronic study of writing practices and their relation to the socio-historical conditions of production and consumption of written texts will contribute to the establishment of business studies as an academic discipline and enhance the credentials of the business community, which indeed has a writing tradition as long-established as that of other disciplines.

References

Primary sources

Anderson, William [3]1873. *Practical Mercantile Correspondence.* Milan: Bernardoni.

Candelari, Romeo 1899. *Corrispondenza commerciale inglese-italiano.* Milan: Sonzogno.

Cann, Teofilo [2]1883. *The Comprehensive Letter-Writer. A Complete Guide to English Correspondence.* Florence: Paggi / Rome: Alinari & Cook / Naples: Nicola Jovene & Co.

Secondary sources

Bargiela-Chiappini, Francesca / Nickerson, Catherine 1999. *Writing Business: Genres, Media and Discourses.* London: Longman.

Barton, David / Hall, Nigel 1999. *Letter Writing as Social Practice.* Amsterdam: Benjamins.

Biber, Douglas *et al.*1999. *Longman Grammar of Spoken and Written English.* London: Longman.

Bijkerk, Annemieke 2004. *Yours sincerely* and *yours affectionately*: On the Origin and Development of Two Positive Politeness Markers. *Journal of Historical Pragmatics* 5/2, 297-311.

Brown, Penelope / Levinson, Stephen 1987. *Politeness: Some Universals in Language Usage*. Cambridge: Cambridge University Press.

Del Lungo Camiciotti, Gabriella 2002. La scrittura epistolare nella didattica dell'inglese: alcuni manuali per il commercio dell'Ottocento. In Minerva, Nadia (ed.), *Per una storia degli insegnamenti linguistici: bilanci e prospettive. Quaderni del Cirsil* 1, 41-46, http://amsacta.cib.unibo.it/archive/00000543/.

Del Lungo Camiciotti, Gabriella 2005. "I perceive, my dear friend, by your letter of the 20th inst that you are decided on entering upon the career of commerce": Nineteenth-Century Business Correspondence. In Gillaerts / Gotti (eds), 125-146.

Del Lungo Camiciotti, Gabriella 2006. "Conduct yourself towards all persons on every occasion with civility and in a wise and prudent manner; this will render you esteemed": Stance Features in Nineteenth-century Business Letters. In Dossena / Fitzmaurice (eds), 153-174.

Dossena, Marina 2004. Towards a Corpus of Nineteenth-century Scottish Correspondence. *Linguistica e Filologia* 18, 195-214.

Dossena, Marina 2006. Stance and Authority in Nineteenth-century Bank Correspondence: A Case Study. In Dossena / Fitzmaurice (eds), 175-192.

Dossena, Marina / Fitzmaurice, Susan (eds) 2006. *Business and Official Correspondence: Historical Investigations*. Bern: Peter Lang.

Douglas, George H. 1985. Business Writing in America in the Nineteenth Century. In Watts / Ide / Ehlich (eds), 125-133.

Douglas, George H. / Hildebrandt, Herbert W. (eds) 1985. *Studies in the History of Business Writing*. Urbana, Ill.: Association for Business Communication.

Dury, Richard 2006. A Corpus of Nineteenth-century Business Correspondence: Methodology of Transcription. In Dossena / Fitzmaurice (eds), 193-205.

Fitzmaurice, Susan 2002. Politeness and Modal Meaning in the Construction of the Humiliative Discourse in Early Eighteenth-century Network of Patron-Client Relationship. *English Language and Linguistics* 6/2, 239-265.

Fitzmaurice, Susan 2003. The Grammar of Stance in Early Eighteenth-century English Epistolary Language. In Leystina, Pepi / Meyer, Charles (eds) *Corpus Analysis. Language Structure and Language Use.* Amsterdam: Rodopi, 107-131.

Gillaerts, Paul / Gotti, Maurizio (eds) 2005. *Genre Variation in Business Letters*. Bern: Peter Lang,

Hiltunen, Risto / Skaffari, Janne (eds) 2003. *Discourse Perspectives on English: Medieval to Modern.* Amsterdam: Benjamins.

Nevala, Minna 2002. Family First: Address and Subscription Formulae in English Family Correspondence from the Fifteenth to the Seventeenth Century. In Taavitsainen / Jucker (eds), 145-176.

Nevala, Minna 2004. Inside and Out: Forms of Address in Seventeenth- and Eighteenth-century Letters. *Journal of Historical Pragmatics* 5/2, 271-296.

Nevalainen, Terttu / Raumolin-Brunberg, Helena (eds) 1996. *Sociolinguistics and Language History: Studies Based on the Corpus of Early English Correspondence*. Amsterdam: Rodopi.

Nevalainen, Terttu / Tanskanen, Sanna-Kaisa (eds) 2004. *Letter Writing*. Special issue of *Journal of Historical Pragmatics* 5/2.

Parsons, Timothy 1999. *The British Imperial Century 1815-1914: A World History Perspective.* Lanham, Md: Rowman / Littlefield.

Price, Richard 1999. *British Society 1680-1880.* Cambridge: Cambridge University Press.

Richardson, Malcolm 1985. Business Writing and the Spread of Literacy in Late Medieval England. In Douglas / Hildebrandt (eds), 1-9.

Sell, Roger D. 1992. Literary Texts and Diachronic Aspects of Politeness. In Watts / Ide / Ehlich (eds), 109-129.

Skaffari, Janne / Peikola, Matti / Carroll, Ruth / Hiltunern, Risto / Wårvik, Brita (eds) 2005. *Opening Windows on Texts and Discourses of the Past.* Amsterdam: Benjamins.

Taavitsainen, Irma / Jucker, Andreas (eds) 2003. *Diachronic Perspectives on Address Terms Systems.* Amsterdam: Benjamins.

Tieken-Boon van Ostade, Ingrid 1999. Of Formulas and Friends: Expressions of Politeness in John Gay's Letters. In Tops, Guy / Devriendt, Betty / Geukens, Steven (eds), *Thinking English*

Grammar. To Honour Xavier Dekeyser. Leuven: Peeters, 99-112.

Watts, Richard J. / Ide, Sachiko / Ehlich, Konrad (eds) 1992. *Politeness in Language: Studies in its History, Theory and Practice.* Berlin: Mouton.

Watts, Richard J. 1992. Linguistic Politeness and Politic Verbal Behaviour: Reconsidering Claims for Universality. In Watts / Ide / Ehlich (eds), 43-69.

MARINA DOSSENA

Forms of Self-representation in Nineteenth-century Business Letters

It is only in very recent times that nineteenth-century business discourse as a variety of domain-specific English appears to have begun to be investigated *per se* (Dossena 2004, 2006; Del Lungo 2005, 2006; Dossena / Fitzmaurice 2006); however, it certainly has great potential to offer valuable insights, not only as far as the history of specialized discourse is concerned, but also in relation to the cultural framework within which it was employed, with its prescriptions, expectations and presuppositions.

The aim of this chapter is to discuss the linguistic strategies employed by encoders[1] of nineteenth-century business letters, in order to introduce themselves and the purpose of their letter. As this relates to issues in asymmetrical power relationships, such strategies inevitably involve positive and negative politeness moves (Brown / Levinson 1987), and encoders may be seen to adopt fluctuating identities (see Goffman 1959: 17), depending on how forcefully they wish to argue their case and – as a result – how powerful an image they wish to convey of themselves.

For this study we shall take into consideration the complete texts currently being transcribed from original manuscript or typescript sources for inclusion in the Corpus of 19th-Century Scottish

1 In our study we prefer the more encompassing term 'encoder' to indicate the source of the document, as business letters are often written by secretaries on behalf of their employers. Each non-holograph letter may therefore have two 'authors': the employer, who dictates the message, and the scribe (the secretary) who conveys that message in the typical phrasing of business correspondence. As it may be impossible to trace the original notes, it would be incorrect to attribute the text exclusively to one or the other. Although this is of course not the case where holographs are concerned, we chose to use just one term anyway for the sake of terminological consistency.

Correspondence (henceforth 19CSC – see Dossena 2004).[2] At the moment 19CSC comprises a total of ca. 100,000 words, 50% of which derives from ca. 250 business letters of varying length, mostly by male encoders;[3] occasionally, letters by female encoders are also found, though they are typically restricted to the field of job applications and letters of reference for prospective employees. The business section of the corpus includes samples from the Bank of Scotland correspondence (55 letters for a total of ca. 7,000 words), letters relating to two families of printers and publishers (the Blackies, based in Glasgow, and the Blackwoods, based in Edinburgh); various law firms, manufacturers and contractors based in Lowland Scotland; and the administration of Glasgow University.[4] In the next stages of corpus compilation the range of sources is clearly expected to increase, in order to include both different sources for the same type of business (e.g. banking), and a wider range of business and geographical locations.

At this stage, given the relatively limited number of words in the corpus, no statistical data may be offered in relation to linguistic findings. However, we may already outline qualitative indications pointing towards potentially interesting developments; for instance, from the point of view of historical pragmatics, observations on formality and politeness markers may already prove of a certain interest, and preliminary studies are already available (see Dossena 2006).[5]

---

2 I gratefully acknowledge permission granted by the Bank of Scotland, the Glasgow University Archive Service, and the Trustees of the National Library of Scotland to access and quote from materials held in their archives. It should be noted that such permission does not extend to third parties. I also wish to thank prof. Richard Dury (University of Bergamo) for his contribution to the compilation and study of 19CSC, and for his comments on an earlier version of this paper.

3 The remaining 50% of the corpus pertains to private correspondence (150 letters, by male and female encoders; this includes emigrants' letters, mostly by male encoders).

4 In the second half of the 1860s the New College at Gilmourhill was being built, and interesting contractors' correspondence is recorded in that respect.

5 Note that this study concerns authentic materials, while the prescriptions of letter-writing manuals for foreign learners are discussed by Del Lungo Camiciotti (2005 and 2006).

In particular, we intend to discuss the ways in which different categories of encoders (of varying social status and levels of schooling) appear to follow politeness maxims (Leech 1983) – especially the maxim of modesty, on account of the need to downtone their status, in order to avoid FTAs – when they present themselves and their business. At the higher end of the social scale, we shall take into consideration publishers and bank officials; at a less powerful level, contractors and secretarial job seekers; and, further down the social scale, letters encoded by partially- or minimally-schooled encoders[6] pleading for financial help. Of course this implies that our definition of 'business communication' is sufficiently broad to encompass all these types of letter: such broadness derives from the belief that categories can only be defined not as *a priori* frameworks, but on the basis of actual content and of the density of linguistic features characterizing such exchanges.

1. Negotiating power and identity in business correspondence

As in present-day interactions in writing, also in nineteenth-century letters a sense of identity is conveyed by the choice of the forms of address employed in the salutation and in the closing greetings; in particular, such choices indicate to what extent the initiator of the exchange wishes to signal social and psychological proximity to the recipient. A 'neutral' salutation like *Dear*... may take on a more personal overtone if it is intensified by superlative forms and/or possessive adjectives (as in [*My*] *Dearest*...). Similarly, closing greetings may be more or less emphatic: business letters typically end with the formula "Yours Most Sincerely" or "I am [...] Your very Obedt Servt", in which, despite the abbreviation,[7] we have the traditionally deferent tone expected in exchanges between strangers.

6 These terms are adopted from Fairman's studies (2000 and 2003).

7 On the form and function of abbreviations in correspondence between Early and Late Modern English see Markus (2006).

Closeness and familiarity, instead, are signalled by formulas like "believe me to be ever most Affectionately Yours", "believe me ever & ever Your Most Affectionate" or "I remain your most sincere Friend".[8]

Of course, business relationships could and did change over time, as the acquaintance between interlocutors became deeper; as a result a more personalized register could develop, in which personal comments could be included in what otherwise would have been expected to be a more detached type of communication. Indeed, the strategies of positive and negative politeness are seen to be employed consistently throughout, not just in relation to the interlocutor, but also in relation to the development of the connection between participants; in (1),[9] for instance, we see the publisher addressing an author with whom the business relationship seems to be making good progress – reference is made to a "new History", implying that other works have already been published, and therefore the address form is more involved and involving than in (2), addressed to a less authoritative recipient, and in which the publisher expresses doubts (i.e., distance) as to the potential success of the solution envisaged:

(1) My Dear Sir Archibald Along with this I enclose our offer for the first edition of your new History, which will I hope prove agreeable to you

(2) My Dear Oliphant [...] With regard to the 4th Ed^{n} of your book we cannot give any exact statement. [...] We very much doubt whether the cutting up the book into parts and selling it at a cheaper rate could bring in anything worth speaking of

Although both come from the same line of business and the authority of the encoder is comparable, we see that significant changes may be dictated by the mutual power relationship existing with the recipient and by the object of the exchange: the encoder chooses to attribute greater or lesser authority to himself also in relation to the (real or perceived) status of the recipient in a specific situation. In what

8 On eighteenth-century usage see Tieken-Boon van Ostade (2003). Opening formulas in the same century are also discussed by Austin (2004).

9 For the sake of brevity, in all quotations word, line and paragraph breaks are omitted, except where this is deemed to be significant. While underlining is kept as in the original, italics show added emphasis.

follows we shall investigate the dynamic picture that emerges from the possibility of these changes in the linguistic strategies adopted by different categories of encoders, as there appears to be a certain range within which these levels of authority may be made to fluctuate.

2. Politeness moves and the power of capital

As mentioned above, the participants' relative status influences the way in which politeness strategies are employed. This is seen very clearly in letters encoded by bank officials, in which requests are encoded more or less directly and more or less forcefully, depending on whether the recipient is an agent, a correspondent or a client. On the other hand, lower-status encoders, such as contractors, address potential customers in more or less deferential terms depending on their level of schooling, while seekers of financial help mitigate their requests with positive politeness moves, which may include references to shared cultural values, such as religious or political creed. As a result, the way in which encoders present themselves appears to follow a complex set of rules dictated by social, in addition to pragmatic, constraints.

2.1. The correspondence of bankers, publishers and landowners

In the subcorpus of bank correspondence analyzed by Dossena (2006), the client is perceived to be superior by definition, although – depending on the recipient's actual status – varying degrees of distance may be encoded through salutation formulas, closing greetings and – within the body of the letter – the use or avoidance of imperatives. For instance, in (3) the request is presented off-record through reference to future necessity ("it will be necessary"), and is preceded by a grounder in the form of a statement ("the order p £ 200 paid to_day […] overdraws the credit established with […]") and the outline of a hypothetical scenario ("Should M^{r} [C] require further pay^{ts}") from which the necessity would derive:

(3) The BofS. beg to intimate to Messrs. […] that the order p £ 200 paid to_day of M^{r} [JC] overdraws the credit established with Coutts +C by £ 50 _ Should M^{r} [C] require further pay^{ts} it will be necessary for Messrs C. to renew the credit

In (4), instead, the request is expressed on behalf of a third-party subject (the Directors): the scribe thus avoids all attribution of power to himself as the actual encoder:

(4) With reference to the Cash Account here in your name the Directors are desirous of receiving additional security for this credit in respect of the death of M^{r} [RHM].

When the relationship between participants is on more equal terms, such as in the case of letters to business partners, we see an interesting combination of matter-of-fact comments and polite requests – (5) and (6) – and personal remarks the function of which is mostly phatic, as shown in the extract of a letter from a publisher to one of his authors (7):

(5) The BofS. shall feel obliged by the Commercial Bank procuring them an opinion of the character + circumstances of M^{r} [WA] of Newburgh.

(6) I beg leave to trouble you for information as to the means + standing of [JHM], Pencaitland Collieries, Tranent with special reference to his sufficiency for £ 200 in the way of business

(7) Your reputation is now so firmly established that a measure of success is a certainty & we have accordingly encreased upon former terms in consideration of the lesser risk we run. That a great success will be obtained we feel hopeful […] We trust that this new undertaking may prove a fresh bond between us & should there be any points in our proposal which you would like altered we will gladly do what we can to meet your wishes. […] I am delighted with the M.S. & augur great things, if the insanity about Free Trade does not interfere with it. That mania however cannot last for ever & at all events we have the numerous party who think with us to rely upon. […] We had as I told you Miss [S] – staying with us who is a most amusing person / I never heard any woman talk so much & so well. It is true that <u>the much</u> rather gets the better of <u>the well</u>

Finally, in letters to agents, perceived to be subordinate by definition, the encoder expresses meaning most directly, whether it is to make requests and signal urgency, or negotiate conflict with an outcome that may be as favourable as possible to the company (with which the encoder clearly identifies). In bank correspondence, for instance, agents are typically addressed as 'Sir' (though a few occurrences of 'My dear Sir' are also recorded) and imperatives occur more frequently than one might perhaps expect in formal communication, if present-day standards were applied, although such imperatives are couched in the polite formula 'be pleased':

(8) your Letter of 7 Curr^{t} is here _ But *be pleased* to Report expressly whether

(9) Sir, *Be pleased* to open an account in the Banks Books at Leith with a Credit of one Thousand Pounds Sterling in name of

Given the asymmetrical power relationship existing between directors and agents, equally forceful and direct is the Directors' way of expressing their displeasure with an agent's conduct; their evaluation is conveyed either expressing surprise at unexpected news, or expressing failure to understand the reasons behind the disapproved behaviour. Disapproval of subordinates' behaviour is also expressed in unmitigated tones by employers who complain to their associates; in (10) the encoder conveys his hope that the situation will improve when the cause for dissatisfaction is removed:

(10) I am glad to hear that Falconer is disposed to come out here, I think him a very suitable man and since I last wrote the impression grows on me more strongly that Taylor must be changed and that without delays. The fact mentioned in my last that he had never posted his ledger nor balanced his cash speaks strongly as to his carelessness and laziness [...] Though Taylor has been here twice he has never balanced with M Clement nor yet instructed him as to keeping his books, for I find his books sadly in deray + behind though he had good notice of my coming.

On the other hand, it is not only face-threatening acts that can be performed bald on record; also references for prospective employees may be given very clearly, in evaluative tones that avoid any ambiguity as to the favourable perception of the encoder. Two examples are given below:

(11) I consider David Thomson an excellent Coachman [...] He is a very tidy servant both in the stable & in his person. seems to have a good knowledge of the care of horses _ and is honest _ sober & attentive.

(12) In compliance with your request Ihave much pleasure in stating that you discharged in a most satisfactory manner the duties of Teller + Accountant at the Agency of the Bank of Scotland at Stonehaven . for several years [...] Ihave every confidence that you would conduct aBank Agency with great prudence and caution . I consider +esteem you a gentleman of high honor and integrity and I am confident that in whatever situation you may be placed . you will add the most upright conduct to the faithful + honest discharge of your duty

Apart from cases in which conventional deference to clients is expressed through the attribution of superior status, when errors are made and apologies are due the encoder may find himself in a position of actually lower authority than his recipient; in the example below we see that the message is personalized by means of a first-person subject indicating the encoder, although the mistake "has been committed" by an unspecified agent; this has the advantage of thematizing the mistake (to which the client's attention is drawn), while obscuring the actual responsibility within the bank:

(13) I am sorry to perceive that an error has been committed in our letter to you of the 1st Inst. The amount of the order on Stirling placed to the credit of your account here is, as you mention, £ 1000 and I beg to return our letter with the necessary correction and my best apoligies for the mistake

The quotation also provides an example of the use of the verb 'to beg', employed to introduce action on the part of the encoder without prior permission on the part of the recipient. As a matter of fact, this is a very common phrase in correspondence, and it typically collocates with 'leave':

(14) We beg leave to refer to your decision

(15) I beg leave to trouble you for information

(16) I beg leave to return you the State relative to

In addition, this verb collocates with other verbs – for instance, encoders 'beg' to inform / say / acquaint / enclose / intimate / return / acknowledge:

(17) I beg to acknowledge receipt of your letter of 19 Inst. + to acquaint you that

(18) I have just received your letter of the 19 Inst and beg to say that it will not be necessary for you to come in to Edin^{r}.

However, the real semantic value of this *captatio benevolentiae* strategy is in fact very low; when the encoder does have lower status than the recipient, deference is stressed by means of references to the "great respect" due to the interlocutor. The following quotations, for instance, are taken from the correspondence addressed to Earl Gower by Patrick Sellar; here the encoder did indeed have the status of a subordinate, so the concluding phrase had to reflect what was conventionally expected in such a social context:[10]

(19) I have the honour to remain # with every respect # My Lord # Your Lordships faithfull Servant

(20) I have the honour to Continue with the greatest respect # My Lord # Your Lordships faithfull humble Servant

(21) I remain with the greatest respect My Lord #Your Lordships faithfull humble Servant

Similarly, social distance is emphasized when there is a metatextual comment on the encoder's supposed presumption in addressing the recipient, and there is even a request for pardon for such a breach of propriety, which in other cases is described as 'taking liberties' (see § 2.2. below):

(22) I hope you will pardon the freedom I have thus used, and # beleive to be, with much respect, # Dear Sir, # Yours very faithfully

10 Note how the line breaks highlight the individual elements in the formula, making this a kind of ornament preceding the signature.

This may be accompanied by positive politeness moves like a comment on the 'goodness' of the recipient on his complying with the request or a reference to the 'smallness' of the request, aimed to maximize approbation of the recipient and minimize cost to the same:

(23) I request you will have the goodness immediately to propose

(24) Your attention to this small request will oblidge me very very much

2.2. *Contractors and job seekers*

In business exchanges, requests may concern the beginning of a new business relationship on the part of prospective employees or prospective contractors. In these cases, the encoder is aware of the threat his act poses to the recipient's positive face – hence the need to express meaning in as acceptable a way as possible. In the quotations below, for instance, the encoder portrays himself as 'taking liberties' – that is, overstepping the limits imposed by his own status – but, at the same time, pre-empting potential criticism through this very same move, and putting himself in a position to perform the face-threatening speech act without waiting for formal permission to do so:

(25) I take the liberty of asking you to keep me in view.

(26) I take the liberty of writing to ask if you could give me any German & English translation.

The encoder may also try to put himself in a favourable light, through references to shared cultural background with the recipient – for instance, this is the case when references to religion are made. The corpus currently comprises two instances, one encoded by a fully-schooled man, the other by a partially-schooled woman, in which explicit reference is made to the Kirk as a metonymy of the encoder's beliefs and values and of the social network in which both the encoder and the job applicant participate:

(27) I have had much experience & in teaching as Tutor in Families of high respectability. have studied eight years in Edin^{r} for the *Church of Scotland* & am on the eve of taking license

(28) we have now got in amongest *a set of good honest presbyterians* who do all that lies in there power to encourage Mary in hir work

In these cases we find an element of personalization aimed at giving an idea of the encoder's social profile. However, there are also cases in which comments are made in relation to the encoder's personal situation with respect to the recipient – in such instances, although the letter does contain a job application, the private remarks of the encoder signal that the distance between participants is significantly reduced. In the following example, politeness maxims are followed throughout: the encoder (a woman) expresses apologies for potential inconvenience to the recipient, and hedges her statements to a very high degree ("I thought / perhaps / you wouldn't"); she also refers to her parents' approval being connected to the identity of the recipient: a very powerful positive politeness move, indicating the esteem in which the recipient is held by the encoder's family:

(29) I hope you will excuse my troubling you, but I thought perhaps you wouldn't mind my writing to you. I thought you might be in the way of knowing of some secretary work, or copying that one could do_ […] I may remark that I have not said anything to Father & Mother about it, […]; however I am sure that if it was anything you had to do with they wouldn't in the least mind my doing it_ […] it would be very kind of you if you should happen to be able to help me, without bother to yourself_

Exchanges between strangers instead are exemplified in the letters sent by prospective contractors wishing to participate in the construction of the 'New College' at Glasgow University in the late 1860s.

Starting from the paratextual level, we observe that encoders may present themselves as an established firm with its own headed paper,[11] or their notes may be simple hand-written documents, with no

11 For instance, one letter is on headed paper with the Prince of Wales' Crown and motto *Ich Dien* in the top left hand corner, while the sender's address, in ornamental lettering, is centred. Another letter, also with ornamental lettering, is on headed paper with an image of the company's premises.

ornaments at all. Similarly, some contractors just sign the message, which is in a different hand, showing that the text has been drawn up by a professional scribe, probably a secretary familiar with the conventions of business discourse; at the other end of the scale, the encoder himself may be seen to have drafted the text in his own minimally-schooled language, as in the example below:

(30) Mosers Mitchall Allaric + Mitchell # Gentlemen # we woud thonk you to forward us Shed for al the departments of the works Conected with the glasgow university Buildings or we wil Cal for them on wedsenday first at your office + Remains yours # Respectfuly

On occasion, they may refer to how they found out about the possibility to apply – that is to say, they mention a newspaper advertisement, either generically, or more specifically, giving the name of the newspaper in which the advertisement was published, so that the publicity given to the recipient's initiative justifies the request for information on the part of the encoder, although participants are strangers to one another:

(31) Gent^{mn} # Having seen the advertisment in todays paper for estimates for Glasgow New University I have to request you to send me a schedule for plumber work _

(32) Gentlemen # In terms of advertisement in the Herald I beg to apply to you for a Schedule for the Plumber work of the New College with the view of my giving in an estimate for the Job_

As regards politeness moves, the use of 'I beg to apply' and 'respectfully request' occur quite frequently in contexts like the ones outlined above, in which encoder's lower status is conveyed; in addition, the expression of thanks and the idea of obligation deriving from the recipient's complying with the encoder's request are often mentioned at the very beginning of the text:

(33) Gentelmen # Would much oblige by favouring us with a Schedule for the Slater work of New _ University.

(34) Gent^{m} # I would thank you to supply me with a Schedule for the plaster work of New College Glasgow, in order that I might give a tender for that department and oblidge # Your Ob^{t} Ser^{t}

(35) Gentlemen # We will feel obliged by your favouring us with a specification for the smith work for the new University Buildings at Gilmorehill trusting you will favour us with a specification

Interestingly, another case in which 'beg to offer' is employed occurs in a very formal letter in which technical details and price indications are provided – that is, in a letter that appears to meet the requirements of formal business dealings very accurately:

(36) To the # Senate of the University, Glasgow # Gentlemen # Tender for Electric conductor # I beg to offer tender for supplying, and erecting Lightning conductors to the new University buildings_ The conductor consists of solid copper top rod (¾ inch diameter) and solid points, screwing into a cop[r] hole, the whole standing 6 feet above [the] building they are designed [to] protect, [...] The same top rod, and points, with main conducting cable, (3/8 inch diameter) supplied, and fitted complete @ one shilling and sixpence 1/6 per foot for length of conductor, including all charges_

Such specificity is meant to convey an idea of thoroughness and serious business conduct on the part of the encoder. Indeed, concerning self-presentation, we see that three encoders provide additional comments on the possibility of providing external references, samples, and ensuring speedy delivery – clearly, such encoders were already aware of the importance of these elements in promotional discourse:

(37) If any reference is required, I beg to name [...] Engineer for the Caledonian Ry Co for whom I am now building those extensive premises at Buchanan Street Station.

(38) I should be happy to submit a sample for inspection

(39) We are in a position to execute any quantity of Seating with despatch.

2.3. "On hamely fare we dine"

A special kind of business letter is that in which encoders address wealthier recipients in times of financial distress – the following example is a particularly touching case, and unfortunately so far I have been unable to trace any answer to it:

(40) Sir You will perceive from the inclosed that I am authorized to sell an old Bible, [...] and the reason [...] is a very Melancholy one, it is shortly these My Father [...] Has been confined in the cannongate Jail for these three Months past for a debt comparatively But Small, yet small as it is we have and still are unable to get him set at liberty, it is now the only and last alternative, in a word that of succeeding in getting the old Bible sold, [...] having heard that you are much in the habit of collecting old Bibles I have taken the liberty of waiting on you, to know if you will give an offer for the old Bible and should you decline Purchasing it, will you have the goodness to point out the proper Quarter where I can get it disposed of to some advantage for the behoof of my unfortunate Father,

The encoder approaches the recipient (a Presbyterian minister) with a paragraph in which reasons are offered for his proposal and with a reference to the fact that he is aware he is 'taking liberties' – although the difficult circumstances in which the request is made make it admissible.

At other times, ministers could be addressed more indirectly by partially-schooled encoders who still wished to make their conditions manifest, but did so through reference to some points of doctrine. In the following quotation, for instance, a weaver addresses the minister showing remarkable competence in religious discourse:

(41) Reverend Sir, One with whome you are an intire stranger; and who is unworthy of your notice, would humbly beg the liberty of interupting you for a few moments in accnowelidging his gratful obligations to you for the benifit which he hopes he has received from your publick minesterations. [...] May I humbly hope, that I, through sovereign grace, am among the 'poor and afflicted people who trust in the name of the Lord'; and that I have been led, both for myself and for the children whom the Lord hath given me, to seek first the kingdom of heaven and the rightiousness thereof. But I see I must not follow Christ and his truths and ways for the loaves and the fishes. The Lord has seen meet, in his wise and holy dispensations, to feed me with the bread of sorrow and adversity and the water of afflictions. _ I have a large, young family; with <u>very slender</u> means for their support. Through serious distressing causes and events my circumstances have been greatly reduced. But the principle cause has been, the great decline of employment and prices in my way of busniss, that of a Weaver, espashely these two years past, and still more so of late. [...] Dear Sir, it is very unlikely that ever I will injoy the honour and happyness of personal intercourse and verbal communation with

> you in this world: but I rejoice in having spiritual communion with you, and with all that in every place love and call on the name of Jesus Christ, both their Lord and ours.

Here the encoder places great emphasis on affect: his 'prayers / humble requests / humble hopes' show that this letter is not really a begging one, i.e. one in which material support is sought explicitly, such as in the case of the pauper letters discussed by Fairman (2003) – it is instead a letter in which thanks are given for spiritual comfort (though possibly in hopes of material help too).

A slightly different case is that in which the recipient is addressed with a direct request, and this is framed between a comment on the political situation (again aimed at establishing common ground with the recipient) and a reference to a former employee of the recipient's (aimed at personalizing the message through the establishment of a connection with the encoder's family). Finally, reference to the pension that was not claimed aims to put a certain degree of pressure on the recipient, showing that the encoder is aware of what rights his late uncle might have had, but refrained from exercising, thus showing consideration for the recipient's higher status:

(42) altho' I am a staunch Conservative and my people before me yet, it was this Goverment that injured me and I apply to them for redress but I am afraid I will make little of it. # The reason I take the liberty of addressing you is that my rent is unpaid, and altho' I have <no> right to expect assistance from you, yet I apply for a little help from you in my distress. # My Uncle the late [...], Compositor, was for a long time in your employment and when he was no longer able for work I had to do my best to support him. [...] we never applied for a pension from your firm.

3. Concluding remarks

In the nineteenth century, like today, business transactions may be assumed to rely on the kind of hierarchical discourse in which the deployment of positive and negative politeness strategies is crucial. In particular, the politeness maxims of tact, modesty and generosity appear to dictate the way in which interlocutors construct their own *personae* in the letters, in order to achieve their illocutionary (i.e., business) intents. In the samples from 19CSC analyzed in this study, the 'perceived' social status of encoders and recipients appears to be a stronger conditioning factor than age, gender and actual status, as encoders may present themselves with more modesty than might be expected, on account of their purpose and the image they wish to convey to their recipients.

The relevance of actual status, however, cannot be neglected, as greater flexibility in linguistic choices appears to correlate with higher social status and the wider network of interactions that it allows. As wealthier encoders are generally seen to interact with a wider range of recipients, both at an equal and at a lower social level, they have greater opportunities to employ a wider range of communicative strategies in business interactions. Lower-class encoders, on the other hand, normally seem to employ politeness strategies conveying recipient superiority, given the more restricted and specific function of their messages (they are typically contractors offering their services to prospective clients, or people applying for jobs or seeking financial help).

In order to test this hypothesis, our analysis restricted itself mainly to requests, in order to see whether the same communicative function was really encoded differently as actual status changed. This allowed us to see that in bank correspondence the client is perceived to be superior by definition, although varying degrees of distance may be encoded; in other areas of business transactions, letters to partners could also convey greater or lesser authority, depending on the success achieved by the business relationship itself – the more successful this is, the greater the proximity between the partners. While the choice of politeness strategies might appear to be a function of purpose (distancing a less than productive business partner, requesting payment of a distinguished client, or reproaching an agent, etc.) and degree of urgency, the participants' varying roles in the exchange also contribute significantly to the definition of politeness standards.

Awareness of the social milieu in which the exchange occurs is thus particularly important for the investigation of business correspondence, especially when its temporal collocation sets it in a specific cultural framework, potentially quite unlike present-day models. The cultural models of the nineteenth century, with their ideals of politeness and propriety, are inevitably reflected in the linguistic strategies employed to convey respectful deference. At all social levels, encoders strive to reinforce the proximity and solidarity required for a successful business exchange, while maintaining the degree of distance required by social decorum. In the case of wealthier encoders, however, conscious expressions of authority may be more transparent. At the same time, it is equally interesting to observe that skilfully manipulative language uses may also occur in letters from the opposite end of the social scale, almost in an attempt to achieve the same degree of communicative effectiveness found in correspondence issued by encoders with a longer writing tradition and more extensive practice, possibly influenced by the *dicta* of letter-writing manuals.

References

Austin, Frances 2004. "Heaving this importunity": The Survival of Opening Formulas in Letters in the Eighteenth and Nineteenth Centuries. *Historical Sociolinguistics and Sociohistorical Linguistics* 4, at http://www.let.leidenuniv.nl /hsl_shl/ (accessed 18.08.2004).

Brown, Penelope / Levinson, Stephen 1987. *Politeness. Some Universals in Language Usage.* Cambridge: Cambridge Univ. Press.

Del Lungo Camiciotti, Gabriella 2005. "I perceive, my dear friend, by your letter of the 20th inst. that you are decided on entering upon the career of commerce": Nineteenth-Century Business Correspondence. In Gillaerts, Paul / Gotti, Maurizio (eds) *Genre Variation in Business Letters.* Bern: Peter Lang, 125-146.

Del Lungo Camiciotti, Gabriella 2006. "Conduct yourself towards all persons on every occasion with civility and in a wise and prudent manner; this will render you esteemed": Stance

Features in Nineteenth-Century Business Letters. In Dossena / Fitzmaurice (eds), 153-174.

Dossena, Marina 2004. Towards a Corpus of Nineteenth-century Scottish Correspondence. *Linguistica e Filologia* 18, 195-214.

Dossena, Marina 2006. Stance and Authority in Nineteenth-century Bank Correspondence: A Case Study. In Dossena / Fitzmaurice (eds), 175-192.

Dossena, Marina / Fitzmaurice, Susan (eds) 2006. *Business and Official Correspondence: Historical Investigations.* Bern: Peter Lang.

Dossena, Marina / Jones, Charles (eds) 2003. *Insights into Late Modern English,* Bern: Peter Lang.

Fairman, Tony 2000. English Pauper Letters 1800-34 and the English Language. In Barton, David / Hall, Nigel (eds), *Letter Writing as a Social Practice.* Amsterdam: Benjamins, 63-82.

Fairman, Tony 2003. Letters of the English Labouring Classes 1800-34 and the English Language. In Dossena / Jones (eds), 265-282.

Goffman, Erving 1959. *The Presentation of Self in Everyday Life.* New York: Doubleday.

Leech, Geoffrey 1983. *Principles of Pragmatics.* London: Longman.

Markus, Manfred 2006. Abbreviations in Early Modern English Correspondence. In Dossena / Fitzmaurice (eds), 107-128.

Tieken-Boon van Ostade, Ingrid 2003. Lowth's Language. In Dossena / Jones (eds), 241-264.

MARTTI MÄKINEN

Dissemination of Textual Material and Second-Generation Corpora: Finding Intertexts in Early English Medical Writing

1. Introduction

1.1. Aims and materials

The aim of this chapter is to exemplify ways in which second-generation corpora can be used for identifying intertextual links between texts. I will first introduce the method used and discuss the relevance and validity of corpus-based study of intertextuality. I will then comment on the phases of the method that could be automated, and will give an example of a program that has been developed for this purpose. The material I am using is mostly medieval, dating from between 1330 and 1550. The chapter will, nevertheless, touch upon topics that are relevant for the study of LSP and intertextuality of non-fictional texts in general. The material-specific issues are mostly related to the criteria used in automated searches, restrictions to the method, and the recognition of textual parallels.

Second-generation corpora are compiled for specific research problems, sometimes focusing on a certain register, whereas general corpora encompass several registers and are designed for multiple purposes.[1] The corpora I have used, *Middle English Medical Texts*

1 First-generation corpora are e.g. the Brown Corpus (see <khnt.hit.uib.no/icame/manuals/brown/INDEX.HTM>), or The Lancaster-Oslo-Bergen Corpus of British English (LOB, see <khnt.hit.uib.no/icame/manuals/lob/INDEX.HTM>); examples of second-generation corpora are the British National Corpus (BNC) and the Michigan Corpus of Academic Spoken English (MICASE, see <www.lsa.umich.edu/eli/micase/index.htm>). As regards historical or diachronic corpora, The Helsinki Corpus (HC, see <khnt.hit.uib.

(MEMT) and *A Corpus of Herbals in English* (ACHE), are both purpose-built corpora, the former having first served as material for the project called 'Scientific Thought-styles: The Evolution of English Medical Writing',[2] and the latter for my PhD dissertation. MEMT contains 495,322 words from 77 manuscripts in 86 text files; the time span of the corpus is from 1375 to1500. ACHE is a corpus of herbal texts, containing approximately 180,000 words in 14 texts, four of which are post-1500. The time span of ACHE is c. 1340-1550.[3]

The vista of medieval medicine that is provided by MEMT and ACHE is filtered through the choices of generations of editors of medical texts, starting in the mid-nineteenth century and continuing to this day (Taavitsainen *et al.* 2005). MEMT contains mostly previously edited texts, although there are few cases where the corpus compilers have chosen to transcribe texts not available in printed editions in order to balance the selection of texts (Taavitsainen *et al.* 2005; Pahta *et al.* 2005).[4] The selection is divided into three traditions of medieval medical writing: Surgical texts, Specialised texts, and Remedies and *materia medica*. The first tradition includes anatomical texts in addition to the surgical texts proper; the second tradition includes texts that are learned in origin, and focus on specific ailments or certain methods of healing, and encyclopedic texts on natural philosophy; the last tradition contains recipe collections, recipes for specific illnesses, health guides, and herbals. In addition, MEMT includes a section on

no/icame/manuals/HC/INDEX.HTM>) was the first multi-purpose corpus to be compiled; historical second-generation corpora are e.g the Corpus of Early English Correspondence Sampler (CEECS, see <khnt.hit.uib.no/icame/manuals/ceecs/INDEX.HTM>) and the Zurich English Newspaper Corpus (ZEN, see <es-zen.unizh.ch/>).

2 'Scientific Thought-styles: The Evolution of English Medical Writing' is a project of the Research Unit for Variation, Contacts and Change in English (VARIENG), at the Dept. of English, University of Helsinki – see <www.eng.helsinki.fi/ varieng/team4/1_4_4_projects.htm>.

3 The reason ACHE continues well into the sixteenth century is the fact that the medieval practices of book production continued strong despite the introduction of the printing press in England in 1476. The authors and printers recirculated the extant material, and it wasn't until c. 1550 that new, original compositions started to appear (Norri 2004: 100).

4 Of the 86 texts or text extracts included, 78 are based on MS editions, six are MS transcriptions by the compilers of MEMT or other present-day scholars, and two are based on facsimiles of early printed books.

medical texts in verse, and an Appendix of trilingual texts from c. 1330-1340. Despite the division of MEMT into traditions, these are not clear-cut and they partly overlap, e.g. a list of recipes may be preceded by a highly theoretical introduction discussing the symptoms and treatment of a particular disease. ACHE holds previously edited pre-1500 texts and early printed material from the first half of the sixteenth century. The texts are all herbal texts, with slight variation in contents: most of the texts are extracts including several herbal entries, which usually begin with the name of the plant, moving on to the description of the plant, its habitat and time of blooming, and finishing off with the medical properties of the plant in question. Texts differing from this general formula are treatises dealing with one plant only, and *synonyma*, which list plant names with equivalents in several languages, thus aiding in the identification of plants, and omitting the rest of a default herbal entry.[5]

1.2. Medieval intertextuality

Intertextuality is a term that is infrequent in combination with words like 'medieval' and 'non-fiction'. The reason I am applying it to the afore-mentioned fields is that Kristeva's idea of texts being mosaics of quotations, or that "every text is the absorption and transformation of other texts" (Schwartz 1996: 13-14; Allen 1999: 39) describes adequately the ideology behind medieval text production, and literally the method of text production. As St. Bonaventure noted in his *In primum librum sententiarum*, even the author[6] of a text relies on the words of others, i.e. nobody writes isolated from the preceding textual

---

5 The physical descriptions of plants in medieval herbals are often very schematic, and give no useful information to the reader as regards identifying them. Therefore, it may have been useful to list the foreign variants of the name of the plant and also the vernacular aliases, confiding that one of them would be known to the reader of the text.

6 St. Bonaventure's full list of writers is 'scribe, compiler, commentator and author'. None of these is said to write solely his own material: thus all these – scribe as a faithful copyist, compiler as a synthesis maker of several texts, commentator as an explainer of others' texts, and author – draw on the previous, canonical tradition of written knowledge.

world. Reliance on the words of others is even further emphasised by the Scholastic reverence for the written word, and the respect of authorities – copyright was an unrecognised concept then.

The medieval book and text production must be seen in the light of the practices of the time and 'institutional' conventions: reverence for the written word and the constraints set by the contemporary intellectual culture not only accepted but also required the extent of text reuse that has been attested in the surviving manuscripts (Bergner 1995: 48). The canonical *auctores* of Antiquity were the measuring rod against which new compositions were evaluated: the *auctores* and their texts were the sign of authority and authenticity (Schoeck 1984: 108-110). Within this framework, I have adopted the term 'non-transgressive intertextuality' to describe the phenomenon of medieval text reuse (Chandrasoma *et al.* 2004: 173), i.e. the policies that governed text production and the conventions of liberal use of unacknowledged sources.

2. Establishing intertextuality: methodological issues

The point of departure in my method is the way in which medical texts/books were produced in the Middle Ages: almost any medieval English medical text was wholly or at least partially indebted to Classical sources, obviously those that had survived and were accessible at the time (see Tyrkkö, this volume). As medieval texts re-circulated the material over and over again, it is reasonable to assume that the field of medieval medical writing in English was laden with textual parallels and intertextual links. Shared knowledge was common knowledge then, and free for anyone to copy and reuse.

The focus in my work is the medieval herbal textual tradition and the textual links between herbals and between herbals and other contemporary medicine texts. ACHE is the primary corpus in the sense that the formulation of the study and method has been dictated by extant and accessible herbal material.

My approach starts with the generation of cue lists of plant names, which I use as keywords for automated searches on the corpora. The plant names need not be the most frequent in the corpora I am using, but the most likely to be found in any text of the corpora, thus a high frequency plant name occurring in only one or two texts would have to make way for a plant name occurring less frequently but with a more even distribution over the corpora. The program packed with MEMT, *MEMT Presenter*, allows data-driven generation of word lists, i.e. I have been able to choose the spelling variants of lexemes denoting plants directly from the processed corpus texts, thus creating lemmas to be used as input lists in the searches.

As my research concentrates on herbals, plant names provide a good starting point, a kind of 'lowest common denominator' in shared text passages. After finding the *loci* where the plant names occur, I have created a sub-corpus of the retrieved text passages, which have then served as input for closer scrutiny. I have focused in on the context of the cues and tried to establish textual parallels, of which most are phrase or sentence length passages.[7]

Another object of study is to identify 'textual parallels' as an indication of intertextuality. The problem with establishing textual parallels is how much variation can be tolerated, or how much in common the passages have to have before they can be pronounced parallels. In a study of recipe variants by Hargreaves (1981: 101-104), the author has chosen to use only narrow textual parallels, i.e. the variation is limited mainly to the organisation and omission of parallel constituents, but the essential elements (ingredients, recipe procedure, the form or recipe) must be present. He lists 40 variants of one recipe, which can be described as a simple[8] based on pellitory that cures head- and toothache. I have included here two recipe variants listed by

7 The fact that in ME texts sentence boundaries are rarely marked by anything as conventional as a full stop sets certain limitations to the definition of searches: I cannot retrieve a 'whole sentence' as there is no way to define an exhaustive list of devices scribes used to distinguish two sentences from each other. However, it is possible to retrieve a whole line (i.e. a sequence between two line breaks), and retrieve *n* characters from either flank of the keyword ($n \geq 0$). This affects the nature of the study: for example, alignment of sentences between corpora or texts in the corpora is not possible.

8 A recipe or medicine including only one (effective) ingredient.

Hargreaves that can also be found in MEMT, and two additional parallels, that are not in Hargreaves (1981):

(1) Ad purgandum caput. take peletur of spayne, and chewe þe rote þre dayes a good quantite, and hit shal purge þe heued &~ do awey þe ache, and festyn þe teeþ in þe gomes.
(MS BL Additional 33996, ff. 80b-1a; Hargreaves 1981: 101, #1)

[For purging the head: take pellitory, and chewe a good quantity of the root for/on three days, and it will purge the head and do away the ache, and fasten the teeth in the gums]

(2) An other Take peletre and chew þ^e rothe iij days and be hole
(MS Medical Society of London 136, f. 1r; Hargreaves 1981: 104, #38)

[Another: take pellitory and chew the root for/on three days, and (you will) be healthy]

(3) For akyng & turnement in þ^e hede Take þ^e rote of peletre & chew yt iij dayes in þ^e mouth & yt schall clense þi hede. & voyde þ^e hakyng & fastyne þi teth.
(MS Trinity College Dublin 158, f. 83r; Benskin 1998: 142)

[For the ache and pain in the head: take the root of pellitory and chew it for/on three days and it will cleanse your head, and relieve the aching and fasten your teeth]

(4) Anoþer [S] clensynge for þe hevede: Tak pellettre & chew þe rott iii days, & it sall clens þi heved & do away þe warkynge, & festyn þi teth
(MS Wellcome Medical Library 225, f. 2v; Jasin 1983: 43)

[For the headache: [...] take pellitory and chew the root for/on three days, and it will cleanse your head and do away the pain and fasten your teeth]

Comparison of the parallels (1-4) reveals that all but one (2)[9] state the purpose, the malady the cure is intended for; they all mention pellitory root as the ingredient; they give chewing of the root for three days as the procedure in order to achieve the desired effect, which is stated in three recipes (the ceasing of head- and toothache). The differences in the extracts are in the choice of lexical items used for the purpose, all of which are synonymous, and the omission of the desired effect in

9 Even in (2) the purpose is given earlier in the MS text ('An other' in the extract, 'Ach off hede' in the MS, four recipes before Harvey's example).

(2). The choice of criteria for the parallels from (1) through (4) is a clear-cut one, and leaves little space for speculation or criticism for the identification of parallels. However, using the narrow definition of a textual parallel, one risks overlooking parallels that are thoroughly paraphrased, even to the extent of changing the syntax and text type.

2.1. Best and worst scenario in parallels

Extracts (1-4) represent the best scenario of finding textual parallels; the worst scenario is a parallel that goes unidentified. Only slightly better are items like (5) and (6):

(5) LITONTRIPON. þurlynge þe ston yt ys vndurstonde. Properlyche yt brekeþ þe stone and putyþ hym ouȝt, ȝef yt be on þe reynes or on þe bleddre.
(MS Glasgow University Library Ferguson 147, f. 28v; Carrillo Linares 1997)

[*LITONTRIPON* is understood as 'piercing the stone'. Characteristically, it breaks and ousts (gall) stones, if they are in the kidneys or the bladder]

(6) Saxifrage. Saxifrage hauyþ his name for-þi þat it brekiþ þe stoon. Some men clepen hit petroselinum.
(MS Stockholm Royal Library X.91, f. 42r ; Frisk 1949: 195)

[*Saxifrage. Saxifrage* has its name because it breaks the stone. Some men call it parsley]

These two extracts would be very easy to find thanks to the vernacular explanation of the Classical names of the herb, as none of the Greek or Latin names (*litontripon, saxifrage, petroselinum*) are shared by the extracts. The contexts of the explanations do not help much in an automated search, nor is *litontripon* a spelling variant of *saxifrage* or *petroselinum* (which is a false identification for *saxifrage*, probably due to similar medicinal properties), thus some botanical knowledge is required in order to collect a complete lemma of synonyms for *saxifrage*. The parallelism here is the knowledge shared about the etymology of the name of the plant; it is akin to the name lists in herbal entries and *Synonyma* texts. Items (7) and (8) are as difficult to identify as (5) and (6):

(7) Item pulvis [of betony] potatus post cenam juvat stomacum et facit digestionem. (MS. Bodleian 483, f. 64r; Grymonprez 1981: 87)

[Also the powder (of betony) drunk after the dinner helps the stomach and promotes digestion]

(8) Powdre of betayne j dronke wiþ water is goud for digestun of blod and hit is goud for þe cowȝe and comfortiþ þe stommak
(MS Wellcome 405, f. 16v; Edmar 1967: 58)

[Powder of betony drunk with water is good for digestion of blood and it is good for the cough and (it) comforts the stomach]

The ingredient is the same, as is the purpose, but little vocabulary is shared between the two. Indeed, the study of medieval English medicine entails studying Latin, Greek and French medicine, at least to some extent. Example (7) is embedded in a ME text with seemingly no reason for the code-switch.

Extracts (9) and (10) illustrate a situation already touched upon, textual parallels not sharing the syntactic structure. This is the case e.g. with recipe paraphrases, which deliver the information content of a recipe but not the form of the instruction (Taavitsainen 2001: 107-108). In herbals, the function of these recipe-like structures is first and foremost to illustrate the medicinal properties of herbs, and only secondarily, if the paraphrase is accurate enough, to instruct the reader to produce the medicine referred to (Mäkinen 2002: 283-284).

(9) A gude drynke for bane of þe hede broken, secundum R. de O. Stamp betoyne wele & drynk þe jus þer-of & lay þe drafe appon þe wounde & it sall brynge a-way þe broken banes & hele þe wond
(MS Lincoln Cathedral A.5.2, f. 312r; Ogden 1938: 76)

[A good drink for the broken bones of the skull, according to R. de O.: crush betony well and drink its juice and lay the (remaining) pulp upon the wound and it will bring the broken bones out and heal the wound]

(10) If it [betony] be stamped and tha~ layde to a wounde in the heed that is smyten with a stroke / it shall hele the wounde fayre / and drawe out the broken bones yf there be ony as the leches do say. (Larkey / Pyles 1941: Bi)

[If it (betony) is crushed and then laid to a wound in the head that has been hit with a stroke, it will heal the wound completely, and draw out the broken bones if there are any, as the doctors say.]

As a matter of fact, most of the retrieved texts in my study are recipes or recipe paraphrases. It is to these that I will refer in order to illustrate in detail the problems in charting intertextual links in my material. Recipe paraphrases have proved to be a problem in the detection of parallels: as shown by (9) and (10), close reading of the material is required in addition to automated searches. These cases are not few, as the forty plant name lemmas I have chosen for my study produce 6000+ hits, and ca. 70-80% of those are recipes or recipe paraphrases. This is not surprising: plants are most often mentioned as recipe ingredients in medieval medical texts. Furthermore, the paraphrases carry valuable information on the relations between texts.[10]

3. Features examined to establish parallels

After automated searches, the qualitative process of close reading allows final decisions to be made on the identification of textual parallels. The method is not unlike plagiarism detection methods, except that as I am working with medieval texts, I have to deal with much spelling variation, and thus the plagiarism or 'text reuse' detection tools intended for modern languages would work with my material only with some adjustment.[11]

---

10 In order to elicit more information from the data, I have used component analysis of the information contained in the established parallels: this approach helps to identify the original genre of the shared text (Mäkinen 2004: 146, 153-154). Recipes are perhaps one of the most formally defined text types, and thus it is also fairly easy to divide them into smaller sub-units or components, each of which carries a certain kind of information. So far I have studied only recipes and their paraphrases; I intend to perform a similar kind of analysis with the rest of the herbal entry constituents, but as they are more genre specific, their analysis might not be as fruitful as with recipes.

11 There are several algorithms for measuring text reuse (see e.g. Schleimer *et al.* 2003), though most of them are based on n-grams. First the texts used for input are normalised, e.g. spaces are filled in and font is changed into all lower case etc. After normalisation, the characters are divided into sequences that overlap with each other. These sequences are called n-grams (or 'clusters'), in which n is a natural number ≥ 2. By overlap I mean that the first

As noted earlier, most of the retrieved data is recipes, thus the description of the close reading process refers to the structures where material-specific explanations are required. The objects of focus in the context of the keyword are:

- nearby noun phrases:[12] synonyms of NPs
  - As many of the retrieved passages are recipes, the keywords often reside in a list of ingredients, thus in close reading one can use the plant name lemmas (which include synonyms of the keywords) as reference material in order to identify the synonyms. This may be carried out by using the retrieved passages as a sub-corpus for searches with the pre-existing lemmas, excluding the keyword lemma.
  - Recipes with the same ingredient lists but different purposes may also be regarded as parallels: in these cases

n-gram starts with the first character in the text, and ends with the n-th character; the second n-gram starts with the second character in the text, and ends with the (n+1)th character. After that the characters in each sequence are translated into a thumbprint number: this is called hashing. The thumbprints of n-grams are then used to calculate a thumbprint for the whole text, be it a full text, sentence, or clause. Then the thumbprints are compared to each other in order to find text parallels.

12 A text reuse measuring tool called 'Crouch', programmed at UCREL (Univ. Centre for Computer Corpus Research on Language, University of Lancaster), uses sequences of words or word stems as n-grams in order to find similar sequences in texts. The program is based on the notion that relations of texts may be studied through the comparison of the sub-units of texts. It needs at least two texts as input, one assumed source and another assumed target text. 'Crouch' aligns each sentence of the target text with each sentence of the source text, and does n-gram comparison at sentence level. The recognition of source-target relations is enhanced by the use of predefined synonym maps and identification of synonymous word pairs by the comparison of the context of words: hence, this approach also covers changes in the choice of lexical items when paraphrasing the reused text (Piao / McEnery 2004; Piao *et al.* forthcoming). The algorithm of the program resembles to some extent the method I am applying, thus a program like 'Crouch' might be helpful in partly automating the qualitative phase of the analysis. 'Crouch' has been tested on EModE material, and it did not fall short of its performance with ModE (Piao *et al.* forthcoming); an extension into ME is conceivable, providing that appropriate lemmas or synonym maps are available.

the common denominator (in addition to the ingredients) can usually be found on the theoretical level, e.g. the recipes in question may be for two diseases both caused by cold.

- nearby verb phrases: synonyms of VPs
  - The treatment of VPs resembles the treatment of NP synonyms, with the exception that there are no pre-existing VP lemmas that we can employ. However, the repertoire of verbs to be used in recipes (which covers most of the material) is quite restricted (cf. Taavitsainen 2001: 99-100), and thus easily covered, although it is less diagnostic than the NPs nearby the keyword.

- components of recipe passages – these are: purpose, ingredients, procedure, administration of medicine, justification, and additional information (such as efficacy phrases) (Stannard 1982: 60-65; cf. Taavitsainen (2001) on variation in the realisation of recipe components). The first four are usually required for an item to be recognised as a recipe, but there are recipes or recipe paraphrases with only purpose and ingredients stated, and the rest left out (relying on the shared knowledge of apothecaries and medical practitioners to cover for it).
  - The ordering of the components may vary significantly between recipes and recipe paraphrases: the order purpose-ingredients-procedure-administration (ABCD) is the default order for a regular recipe or recipe paraphrase, whereas BCDA is the expected order of elements in herbal recipes and paraphrases.[13] Acknowledging the

13 This is due to a textual strategy in herbals: herbals are manuals of plants and their medicinal properties, and these can be illustrated through recipes. When a recipe appears in a herbal entry, usually towards the end, it is primarily an example of the medical use of the plant. The order of the exemplifying recipe, or recipe paraphrase, follows the order 'old information-new information': the name of the plant is already known, thus the recipe starts with it. This particular part of the analysis does not help to identify the parallels, but it tells us something about the relations between texts and the nature of the information shared.

possibility of the rearrangement of the recipe components naturally leads to accepting a wider definition of textual parallels; cf. Hargreaves (1981).

The plagiarism detection tool principles, and the text reuse measuring tool 'Crouch' (see fn. 11 and 12), would be efficient in mapping intertextual links between ME texts, and finding textual links and relations previously undetected. The use of n-grams and a synonym finder (i.e. the comparison of contexts) in 'Crouch' resembles the way in which I detect parallels in my method. This part of the method could be automated if needed. However, variation in spelling and the extent of paraphrases would require extensive lemmas of the lexemes, or synonym maps covering the pivotal vocabulary of medieval medicine.

4. Concluding remarks

This chapter has dealt with the issue of corpus-aided studies of intertextuality in the light of examples from medieval English medicine, which is characterised by a multitude of intertextual links between texts. The methodological innovations that corpora provide culminate in the fact that one can compare any sub-unit of any text in any corpus with any other sub-unit. This makes all parts of a text equal with each other, and the study is not affected by the researcher's bias towards this or that text or part of text. Therefore it is possible to find and map textual relations and intertextual links where they might be unexpected. As regards the material used for this study, it means finding textual variants previously unlisted (e.g. in Hargreaves' recipes), links between texts previously unlinked, and maybe instances of formulaic use in places that do not seem formulaic when studying one text at a time. Knowledge that can be inferred from the established parallels may be revealing about the origin of the shared knowledge (cf. fn. 13). Parallels in texts also frame the unparalleled sections of text, highlighting the areas that were open for the scribe's *ad hoc* writing.

The new prospects offered by second-generation corpora are greater than the glimpse I have just provided here. The need for processed genre- or register-specific information is tremendous and is growing all the time, one of the most obvious academic applications being plagiarism detection. The tools available or in preparation still have their limitations, but already, as they are, they are a great asset to the language research community.

References

Allen, Graham 1999. *Intertextuality*. London: Routledge.

Benskin, Michael (ed.) 1998. Texts from an English Township in Late Mediaeval Ireland. *Collegium Medievale* 10, 91-173.

Bergner, Heinz 1995. The Openness of Medieval Texts. In Jucker, Andreas (ed.) *Historical Pragmatics: Pragmatic Developments in the History of English*. Amsterdam: Benjamins, 37-54.

Carrillo Linares, Maria José 1997. *Edicíon de una version en ingles medio del* Antidotarium Nicholai. Unpublished Doctoral Diss. University of Seville.

Chandrasoma, Ranamukalage / Thompson, Celia / Pennycock, Alastair 2004. Beyond Plagiarism: Transgressive and Non-transgressive Intertextuality. *Journal of Language Identity and Education* 3/3, 171-193.

Edmar, Désirée 1967. MS. Wellcome 405: A Middle English Leech-Book. Unpublished Licentiate Thesis. University of Stockholm.

Frisk, Gösta (ed.) 1949. *A Middle English Translation of* Macer Floridus de Viribus Herbarum. Uppsala: Uppsala Univ. Press.

Grymonprez, Pol (ed.) 1981. "Here Men May Se The Vertues Off Herbes": A Middle English Herbal, c. 1450, MS. Bodl. 483, ff. 57r-67v. (Scripta 3). Brussels: Omirel.

Hargreaves, Henry 1981. Some Problems in Indexing Middle English Recipes. In Edwards, Anthony S.G. / Pearsall, Derek (eds)

Middle English Prose: Essays on Bibliographical Problems. New York: Garland, 91-113.

Jasin, Joanne 1983. A Critical Edition of the Middle English *Liber Uricrisiarum* in Wellcome MS 225. Unpublished PhD Diss. Tulane University.

Larkey, S.V. / Pyles, T. (eds) 1941. *An Herbal (1525)*. Reprint and translation of Banckes Herbal, 1525 (London). New York: Scholars' Facsimiles & Reprints.

McIntosh, Angus 1963. A New Approach to Middle English Dialectology. *English Studies* 44, 1-11; repr. in Laing, Margaret (ed.) 1989. *Middle English Dialectology: Essays on some Principles and Problems by Angus McIntosh, M. L. Samuels and Margaret Laing*. Aberdeen: Aberdeen Univ. Press, 22-31.

Mäkinen, Martti 2002. 'Virtual recipes' – Virtues as a Text Type in Early Herbals. In Raumolin-Brunberg, Helena / Nevala, Minna / Nurmi, Arja / Rissanen, Matti (eds) *Variation Past and Present: VARIENG Studies on English for Terttu Nevalainen*. Helsinki: Société Néophilologique, 271-286.

Mäkinen, Martti 2004. Herbal Recipes and Recipes in Herbals – Intertextuality in Early English Medical Writing. In Taavitsainen, Irma / Pahta, Päivi (eds) *Medical and Scientific Writing in Late Medieval English*. Cambridge: Cambridge University Press, 144-173.

Norri, Juhani 2004. Entrances and Exits in English Medical Vocabulary, 1400-1550. In Taavitsainen, Irma / Pahta, Päivi (eds) *Medical and Scientific Writing in Late Medieval English*. Cambridge: Cambridge University Press, 100-143.

Ogden, M.G. (ed.) 1938. *The* Liber de diversis medicinis, *in the Thornton Manuscript* (*MS. Lincoln Cathedral* A.5.2). London: Oxford University Press.

Pahta, Päivi / Taavitsainen, Irma / Mäkinen, Martti 2005. Catalogue of MEMT Texts. In Taavitsainen, Irma / Pahta, Päivi / Mäkinen, Martti (eds) *Middle English Medical Texts*. Amsterdam: Benjamins.

Piao, Scott S.L. / McEnery, Tony 2004. A Tool for Text Comparison. In Archer, Dawn / Rayson, Paul / Wilson, Andrew / McEnery, Tony (eds) *Proceedings of the Corpus Linguistics 2003 Conference*. Lancaster University: UCREL, 637-646.

Piao, Scott S.L. / McEnery, Tony / Seymour, Michael forthcoming. A Tool for Exploring Text Reuse in Early Modern English Journalism.

Schleimer, Saul / Wilkerson, Daniel S. / Aiken, Alex 2003. Winnowing: Local Algorithms for Document Fingerprinting. In Halevy, Alon Y. / Ives, Zachary G. / Doan, AnHai (eds) *Proceedings of the 2003 ACM SIGMOD International Conference on Management of Data, San Diego, California, USA, June 9-12, 2003*. San Diego: ACM, 76-85.

Schoeck, Richard J. 1984. *Intertextuality and Renaissance Texts*. Bamberg: Kaiser.

Schwartz, Debora B. 1996. The Horseman Before the Cart: Intertextual Theory and the Chevalier de la Charrette. *Arthuriana* 6/2, 11-27.

Taavitsainen, Irma 2001. Middle English Recipes: Genre Characteristics, Text Type Features and Underlying Traditions of Writing. *Journal of Historical Pragmatics* 2, 85-113.

Taavitsainen, Irma / Pahta, Päivi / Mäkinen, Martti 2005. Middle English Medical Texts. In Taavitsainen, Irma / Pahta, Päivi / Mäkinen, Martti (eds) *Middle English Medical Texts*. Amsterdam: Benjamins.

PÄIVI PAHTA

Ful Holsum and Profetable for the Bodi: A Corpus Study of Amplifiers in Medieval English Medical Texts*

1. Introduction

Amplifiers are degree adverbs that scale up meaning – words like PDE *very*, *highly* and *totally*. Their key characteristics, versatility, expressivity and colour, and their capacity for renewal have made them an object of numerous studies. Many studies (e.g. Labov 1984; Partington 1993; Paradis 1997; Ito / Tagliamonte 2003) show that amplifiers have a significant role in social and emotional expression of speakers. This makes them interesting for social, pragmatic and cognitive approaches to language. They are also claimed to be the items that show the most rapid and the most interesting semantic developments in linguistic change (e.g. Quirk *et al.* 1985: 590; Peters 1994: 269). Thus, they provide fruitful material for research on linguistic processes like grammaticalization and de-lexicalization (see e.g. Lorenz 2002; Méndez-Naya 2003).

My interest in amplifiers lies in their register-specific nature and stems from their special character in PDE scientific discourse. I am particularly interested in their rhetorical and persuasive uses in scientific writing: in the ways in which authors of scientific texts use them to justify their choices, convince their audiences, and negotiate

* The research reported here, carried out during my fellowship at the Helsinki Collegium for Advanced Studies, is part of the Research Unit for Variation, Contacts and Change in English, funded in the National Centre of Excellence programme by the Academy of Finland and the Universities of Helsinki and Jyväskylä. The study is also part of my joint project with Irma Taavitsainen on Scientific Thought-styles.

knowledge claims. At the same time, I am also concerned with their other functions in scientific discourse, including their capacity to foreground and highlight distinctions between objects, conditions and processes or their characteristics. My long-term plan is to map variation in the use of amplifiers in different kinds of scientific texts aimed at different kinds of audiences, tracing changes in a long diachrony. In the disciplinary frame, I will focus on the domain of medicine, where a continuum of scientific texts in English extends from the Middle Ages to the present day (see e.g. Pahta / Taavitsainen 2004).

In this study, I examine amplifiers in medieval English medical writing.[1] The study is based on the recently published electronic collection of *Middle English Medical Texts* (Taavitsainen / Pahta / Mäkinen 2005), which facilitates the use of corpus linguistic methods in studying the special language of medicine in a historical perspective. I will focus on amplifiers in adjective modification, which is their prototypical syntactic function in PDE. My main aim here is to provide an overview of amplifiers in this function in the material, and to highlight and illustrate some of the most prominent tendencies in their use. To this end, I combine results from contextualized quantitative and qualitative analyses of the corpus material.

The structure of the article is as follows. In Section 2, I discuss amplifiers as a category and a communicative device, providing some information about their general history and use on the basis of earlier research. In Section 3, I introduce my material and method. Section 4 presents the main quantitative findings on the repertoire and incidence of amplifiers in the material. Section 5 discusses the distribution of amplifiers across the varieties of medical writing. In Section 6, I examine amplifiers in domain-specific discourse. Finally, Section 7 contains some concluding remarks on the results.

---

1 Studies on sixteenth-century and late twentieth-century medical writing are forthcoming (Pahta forthcoming a and b).

2. Amplifiers

2.1. Category

The term 'amplifier' has been used in various senses in earlier research. My definition is based on the *Longman Grammar of Written and Spoken English* by Biber *et al.* (1999: 554-555), who use the term to denote degree adverbs that increase intensity.[2] They point out that some amplifiers indicate degrees on a scale. These so-called boosters include items like *very*, *so* and *extremely*. Other amplifiers indicate an endpoint on a scale. They are called maximizers and include words like *absolutely* and *totally*.

As a functional category, amplifiers are an open class. This means that there is no pre-set finite list of items that could be used to study the phenomenon in any period of English. As Bolinger (1972: 23) points out, "[a]ny list of intensifiers has to be viewed as a sampling rather than a catalog, [...] because [the set of intensifiers] is too open-ended". The range of items included in different studies varies. Many studies are based on a pre-set group of items chosen from earlier literature on the topic. In this study, I have also adopted this method, and focus on a set of amplifiers selected from earlier historical studies of the category (see Section 3).[3]

In PDE, the prototypical syntactic function of amplifiers is adjective modification. The same items can also scale up other adverbs, verbs, determiners, pronouns or prepositional phrases, while

---

2 For Biber *et al.* (1999), 'amplifier' is an alternative term to 'intensifier', a term that has also been used in many different senses. For Quirk *et al.* (1985), 'amplifier' is a subcategory of 'intensifiers', scaling upwards from an assumed norm, while items in the other subcategory, 'downtoners', have a lowering effect, and usually scale downwards from an assumed norm (1985: 589-590). Other terms that have been used in the literature include e.g 'degree modifier' (Paradis 1997) and 'intensifying adverb' (Mustanoja 1960). For a discussion of the terminology, see e.g. Méndez-Naya (2003: 373-374).

3 An alternative, data-driven method is adopted e.g. by Lorenz (1999), who in his corpus-based study of intensifiers as a functional class in PDE has used 'bottom-up' methodology and compiled for his study purposes a list of all expressions by which writers in his corpus modify adjectival qualities in degree.

some of them can also function on the clause-level as adverbials. For example, in Bäcklund's large-scale quantitative study of degree adverbs in contemporary British and American English, 72% of all intensifiers occurred with adjectival heads (1973: 279; see also Allerton 1987: 16). Most later studies concentrate on amplifiers occurring with adjectival heads (e.g. Paradis 1997; Lorenz 1999; Ito / Tagliamonte 2003). In this study, I have also restricted my attention to this group. There are, however, some theoretical concerns in defining what constitutes an 'amplifier occurring with an adjectival head'. The ambiguity arises from the fuzziness of grammatical categories. Two well-known fuzzy boundaries, too complex to discuss here in detail, are especially relevant. The first one is the boundary between adjectives and adverbs, which in ME is particularly fluid (for a discussion see e.g. Mustanoja 1960: 314-315; Nevalainen 1997). The other gradient borderline is that between adjectives and participles (for a discussion, see e.g. Pullum / Huddleston 2002: 78-81; see also Mustanoja 1960: 551).[4] My decisions concerning these ambiguities in the empirical analysis are explained in Section 3.

2.2. Communicative device

As a communicative device, amplifiers are a means of intensification. In the communicative process they are multifunctional, "a vehicle for impressing, praising, persuading, insulting, and generally influencing the listener's reception of the message" (Partington 1993: 178). They convey an interpersonal message in ideational statements (Lorenz 1999: 23). They highlight distinctions, at the same time signalling the author's involvement, commitment and point of view, including truth and value judgements. They are a rhetorical device by which writers can boost their claims and try to persuade their readers of the importance or validity of their observations and recommendations.

Intensification has been characterized as a field of highly individual preference and self-expression (Labov 1984; Lorenz 1999: 24). It is also a culture-specific and socially embedded area of

4 Several theoretical studies discussing the fuzziness of these categorizations are also included in Aarts *et al.* (2004).

language use, and changes in amplifier use are an indicator of shifting norms and practices in a speech community (Ito / Tagliamonte 2003). Thus, amplifiers at the same time act as linguistic cues to the speaker's identity and group membership. They are socially situated and, consequently, register-specific.

2.3. Register variation and the language of science and medicine

In Present-day scientific English, especially in scientific writing, the status of amplifiers is ambiguous. Because of their emotional colouring and associations with colloquial usage and non-standard varieties, academic writers shun them. Guides to academic writing advise against their use, encouraging other, more subtle linguistic means of focusing and boosting instead.

Despite this prescriptive attitude, present-day scientific texts do contain amplifiers. However, their use in scientific writing is different from that in other registers. The corpus-based findings in the *Longman Grammar of Written and Spoken English* show great variation in the repertoire and incidence of amplifiers between the PDE registers of academic prose, news, fiction and conversation. For example, amplifiers like *so* and *really*, common in everyday spoken language, are rarely used in academic prose, whereas items like *extremely* and *highly*, not frequent in other registers, are more common in academic writing (see Biber *et al.* 1999: 564-566). Similar results were obtained in an earlier study on academic prose and conversation by Chafe (1987). Academic registers also vary: a study of evaluative adjectives in the Michigan Corpus of Academic Spoken English (MICASE) shows that amplifier use in academic speech is closer to everyday conversation and casual speech than to academic writing (Swales / Burke 2003). A study of amplifiers in *Medicor*, a corpus of late twentieth-century American English medical writing from six genres, indicates further dimensions of variation, as medical genres like research articles, textbooks and guidebooks show different tendencies in amplifier use (Pahta forthcoming b). Differences between disciplinary discourses are also in evidence (Hyland 2000).

2.4. Diachrony

Because of their emotional functions, amplifiers are especially prone to renewal (Hopper / Traugott 2003: 122). In frequent use, they lose their intensifying force and become inadequate. As a result of "fevered invention" (Bolinger 1972: 18) and recycling, new amplifiers come to use and replace the old ones for the sake of expressivity.

The diachronic shifts in the popularity of the most common amplifiers over time are summarized in Ito / Tagliamonte (2003: 259-262) and Tagliamonte / Roberts (2005: 281-282) on the basis of Mustanoja's discussion of intensifying adverbs (1960: 316-328). The most common booster in OE and EME was *swiþe*. It was gradually replaced by *well* and *full* during the thirteenth and fourteenth centuries. *Full* became the dominant variant, and was surpassed in popularity by *right* in the fifteenth century. In the sixteenth century, *right* competed with *very*, which eventually outshone it. *Really* appeared as the rival of *very* in the eighteenth century. Both items have retained their popularity, but with different profiles. Apart from the most popular amplifiers, a multitude of other items can occur in this function in any period of English. For example, for the ME period, Mustanoja (1960: 316-328) lists 35 "intensifying adverbs" (see Section 3, Figure 2).

The history of the early rivals *swiþe*, *well*, *full* and *right* has recently been studied by Méndez-Naya (2003, 2004, 2005) on the basis of the *Helsinki Corpus of English Texts*, a diachronic multi-genre corpus containing over 1.5 million words of running text from c. 700 to 1710.[5] Méndez-Naya has been able to provide more details concerning the timing of the shifts and the collocational qualities of the main items. The relationship of suffixed and suffixless intensifiers is discussed by Nevalainen (1997) in her study on adverb derivation in the LME and EModE subsections of the *Helsinki Corpus*. Her findings show that the dual forms, such as *full* and *fully*, have different functional patterns, so that, for example, zero forms tend to occur with adjective and adverb heads, whereas *-ly* forms are preferred with verbs and participles (Nevalainen 1997: 166). Other research focusing on individual amplifiers or their use in individual genres in ME includes

5 For information on the *Helsinki Corpus*, see e.g. Rissanen *et al.* (1993).

a study by Peters (1994) on intensifiers in early letters, where he identifies several items not mentioned in earlier literature, and a study on *very* by Gonzáles-Díaz / Adamson (2004).[6]

In the history of English as the language of science and medicine, intensification is an uncharted territory. It is not known, for example, whether the scientific register in earlier periods of English had its own distinct conventions in amplifier use, setting it apart from other registers, or how well the idea of constant, rapid change applies to scientific writing, where conventions change slowly. In view of this, it will be interesting to see how the use of amplifiers has evolved in domain-specific discourse and in different varieties of medical writing across time.

3. Material and method

The corpus of *Middle English Medical Texts* (MEMT; Taavitsainen / Pahta / Mäkinen 2005), on which my study is based, is a computer-readable collection of texts compiled at the University of Helsinki.[7] MEMT contains about half a million words of running text within the domain of medicine from c. 1375 to c. 1500, with a small appendix of texts from c. 1330.

Figure 1 shows the structure and contents of MEMT in a nutshell. The prose material that forms the main part of the corpus is divided into three subsections: surgical texts, specialized texts, and remedies and *materia medica*. This tripartite division reflects and redefines an established categorization of ME medical writings into

6 Forthcoming research includes studies from a workshop on 'English Intensifiers and Grammaticalization', convened by Belén Méndez-Naya in a conference on New Reflections on Grammaticalization, Santiago de Compostela in 2005.

7 For a detailed description of the corpus, see Introduction on the MEMT CD-ROM, or Taavitsainen / Pahta / Mäkinen (2006). The corpus was designed as research material for our project on Scientific Thought-styles. The other project members are Turo Hiltunen, Maura Ratia, Carla Suhr and Jukka Tyrkkö. See also Mäkinen and Tyrkkö (this volume).

academic texts, surgical texts and remedybooks (see e.g. Voigts 1984, 1989; Norri 1992; Taavitsainen / Pahta 1998). Surgical and specialized texts have their origins in the learned traditions of Greco-Roman medical writing, and the vernacular versions may have been aimed at medical practitioners. Many of the surgical texts were originally compiled as university textbooks and translated into English in the course of the fourteenth and fifteenth centuries. Specialized texts, which are also translations, include a variety of sophisticated treatises on physiology and texts focusing on particular ailments or methods of diagnosis and treatment. Remedies and *materia medica* texts represent materials for a heterogeneous audience. They contain various types of prognosticatory texts, recipes and practical advice, herbals, charms, and medical rules written down in commonplace books for private use. In addition to the three main subsections, there is a separate category of verse texts, mainly remedies, where the verse form acts as a mnemonic device, and a small appendix of early trilingual texts, containing remedy material in Latin, English and French.

- o Time span: c. 1375 – c. 1500 (c. 1330)
- o Size: c. 500,000 words (495,322)
- o Contents and structure:
- Surgical texts – 136,527 words (surgical manuals, anatomical descriptions)
- Specialized texts – 87,259 words (physiology; bloodletting, ophthalmology, embryology, urinoscopy, gynecology, plague; astrological theory; learned encyclopedias)
- Remedies and *materia medica* – 216,270 words (recipes, prognostications, charms, dietaries, health guides; herbals, alchemical waters)
- Verse – 31,797 words (remedy material in verse form)
- Appendix – 23,469 words (trilingual remedy material from c. 1330)

Figure 1. The corpus of *Middle English Medical Texts* (Taavitsainen / Pahta / Mäkinen 2005).

The data for analysis was retrieved from the corpus using MEMT Presenter, a flexible search engine specifically designed to accompany the corpus by Raymond Hickey (Essen University). In corpus-linguistic terms, my method is 'top-down'. I compiled a search list consisting of closed-class and open-class items established in earlier studies on intensifiers in the ME and EModE periods. The search list is based on the intensifying adverbs in Mustanoja (1960: 316-330). I added some items to the list from a study of degree adverbs in early letters by Peters (1994) and from my own knowledge of amplifiers in EModE medical texts (Pahta forthcoming a). The final search list contained 57 items and their variants, indicated in their modernized spelling in Figure 2.

- *all, als(o, as), clean(ly), downright, enough, entire(ly), fair(ly), fast(e), fele, ferly, fine(ly), full(y), great(ly), hard(e), huge(ly), inly, much, never so, out, over, perfect(ly), passing(ly), pure(ly), quite(ly), right, so, sore, stark, swithe(ly), throughout (thwertout), utterly, very (verily), well, whole (wholly), wonder(ly)* (Mustanoja 1960)
- *excessive(ly), fervent(ly), hearti(ly), horribly, ineffably, marvellous(ly), most, perfect(ly), special(ly), strong(ly), true (truly)* (Peters 1994)
- *abundant(ly), deep(ly), exceeding(ly), excellent(ly), extraordinari(ly), extreme(ly), high(ly), infinite(ly), splendid(ly), superb(ly)* (Pahta forthcoming a)

Figure 2. Search items and their sources.

The actual corpus searches were designed so that all spelling variants could be retrieved. I do not claim that I have managed to glean every single type and token of amplifiers in the data, but I am convinced that I have caught the vast majority of them.

In this study, I only deal with amplifiers occurring with adjectival heads; these are illustrated in examples (1)-(3) from the data. The same items occurring in other syntactic functions are ignored.

(1) And this evyle othirwhile it is littill, othir while it is muche. And oþer while it is *right* evile and *right* gret. (Lanfranc, *Chirurgia parva*, pp. 91-92)

(2) And alwey at yeven put yn hys ey of oure colerye callyd colerium Alexandrinum [...], tyl he be *perfytely* hoole.
(Benvenutus Grassus, *De probatissima arte oculorum*, p. 70)

(3) in þis season encrecith rede coler, and is like to fire which is *excedingly* hote and drye. (*Regimen sanitatis*, p. 7)

The material contains ambiguous cases, where more than one interpretation of the syntactic roles of the constituents is possible (see Section 2.1). In the analysis, I have decided to include all instances where it is possible to interpret two consecutive words as an amplifier/adjective construction. Thus, examples like (4), where the italicized words can be interpreted as co-ordinate adjectives or an amplifier/adjective construction, have been included in the analysis.

(4) And *grete boistous* metis þat wole not be defied liȝtly, þou shalt not ete.
(Gilbertus Anglicus, *Compendium medicinae*, p. 8)

At the same time, I have decided to exclude all constructions where amplifiers modify participles, or "verbal adjectives" (Mustanoja 1960: 551), including "participial adjectives" (Pullum / Huddleston 2002: 78-81), being aware that this will restrict the scope of generalizations that can be made about amplifier use in the data.

4. Overview: Repertoire and frequency

Of the 57 search items, 28 occur as amplifiers with an adjectival head in MEMT. The items are listed in alphabetical order in Figure 3.

all, als(o, as), clean(ly), enough, exceedingly, excessively, fair, fine, full(y), great(ly), hard, high, hugely, marvellously, much, never so, over, perfectly, passing, pure(ly), right, so, specially, stark, utterly, very (verily), well, wonder(ly)

Figure 3. Amplifiers occurring with an adjectival head in MEMT.

The frequency ranking of the most common items is shown in Table 1. The total number of occurrences is 627 and the normalized frequency for the whole corpus is 12.7 adjectival amplifiers per 10,000 words. Assuming that my initial search list contained all potential lexemes that occur in this function in the corpus, an adjectival amplifier occurs approximately every 790 words. This shows that the construction is not a high-frequency linguistic feature in the material.

Amplifier	**N**	**N/10,000**	**%**
full(y)	183	3.7	29.2
right	109	2.2	17.4
so[8]	88	1.8	14.0
all	74	1.5	11.8
well	43	0.9	6.9
much	34	0.7	5.4
over	20	0.4	3.2
wonder(ly)	13	0.3	2.1
other (19)	63	1.3	10.0
Total	**627**	**12.7**	**100**

Table 1. Amplifiers occurring with an adjectival head in MEMT. N = frequency, N/10,000 = frequency per 10,000 words.

The frequencies of individual amplifiers occurring with an adjectival head vary greatly in MEMT. The most frequent item is *full*. Together with its rare variant *fully*, it occurs 183 times, with the incidence of 3.7 occurrences per 10,000 words. In the frequency ranking it is followed by *right*, which occurs 109 times (2.2 / 10,000 words). As Table 1 shows, the frequencies decline rapidly. The two most frequent items are followed by *so*, *all*, *well* and *much* with frequencies ranging from 88 to 34 (1.8 to 0.7 / 10,000 words). Examples of the six most common amplifiers are given in (5)-(10).

(5) Alsa þe same uryn in an acue is *foull* perilous
(Daniel, *Liber uricrisiarum 2*, p. 147)

(6) & if þu hayf ned to bled, bled on þe xxviij day, for þat is *ryght* gude
(*Diet and bloodletting*, p. 257)

(7) nature ys agreued so moche that sche fylle not her greuance, for yt ys *so* feruent (Canutus, *Plague treatise*, p. 272)

(8) Take þe ballokes of saturion *al* grene (*Antidotarium Nicholai*, f. 11r)

8 The figures for *so* also include the few instances of *never so*.

(9) After boile þam long at þe fire vnto þe oile be *wele* grene; after cole it and kepe it: þis oyntment is *ryȝt* mitigatiue of akyng of apostemeȝ
(Arderne, *Fistula*, p. 12)

(10) ȝyf hys egestioun be opon þe quantite of hys mete, & not blak ne *mykyll* swart in color (Daniel, *Liber uricrisiarum 1*, p. 142)

Most amplifiers have low frequencies in MEMT, occurring less than 20 times in the half-a-million-word corpus. For example, *very*, the predominant booster from the sixteenth century onwards, only occurs seven times. Many amplifiers occur just once. Examples (11)-(15) illustrate some of the low-frequency amplifiers in the data.

(11) And let hym beware of contrarius metys, and for cowes fleshe, ghetys fleshe, and eelys, and such oþer, and specyally of rawe onyons, for they be *specyally* contrarius, as I have prouyd be experyence.
(Benvenutus Grassus, *De probatissima arte oculorum*, p. 57)

(12) The nettle is in greek y-clepid acaliphe. ȝhe is *hugely* hoote, and þer-fore she is clepid vrtica in latin (Macer, *De viribus herbarum*, p. 64)

(13) Then folowith a season callid Estas or summir, and is commonly *excessifly* hote and drie. (*Regimen sanitatis*, p. 7)

(14) putte þe poudre of antymony þere-ynne and sette hit ouer a softe fire til þe vin aigre wexe *verry* rede. (Rupescissa, *Quintessence*, p. 191)

(15) And had I noȝt a-youyn þe warnyng at þe fyrst tyme þat thow haddyst aloyne by her, þu hadist bene *starke* dede.
(De Caritate, *þe priuyte of priuyteis*, p. 142)

5. Distribution across varieties

The repertoire of amplifiers varies from one subsection of MEMT to another. The widest range is attested in the category of remedies and *materia medica*: 25 different items occur in texts in this subsection. Texts derived from the learned traditions have a more restricted range: 14 different items occur in surgical texts and 16 in specialized treatises. The other two subsections have fewer types, which may simply reflect the smaller amount of data in these categories, but may also depend on the distinct nature of the material in comparison with the main corpus. Nine different amplifiers are attested in the verse texts and only six in the early fourteenth-century trilingual texts in the appendix.

The overall frequency of amplifiers also varies from one category to another. This is illustrated in Figure 4. There are fewer amplifiers in the learned traditions. Surgical texts contain 9.8 amplifiers per 10,000 words, specialized texts 12, and remedy texts have 13.6 amplifiers for every 10,000 words. The verse texts have close to 25 amplifiers per 10,000 words.

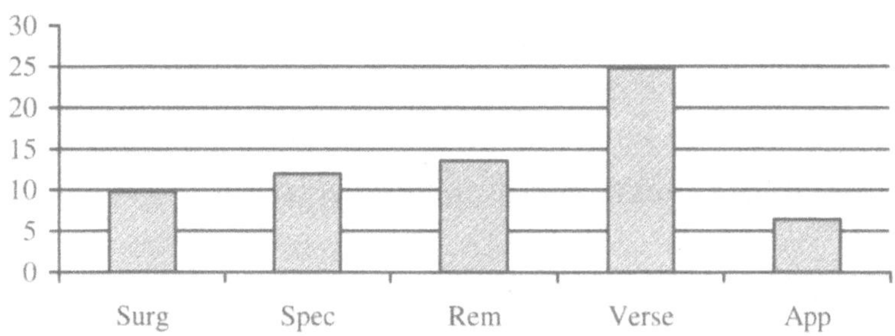

Figure 4. Frequency of amplifiers in subsections of MEMT; normalized figures per 10,000 words.

The three common amplifiers identified in earlier studies as rivals in this period, *full*, *right* and *well*, have different distributions in the data; these are shown in Figure 5. *Full* and *right* are more frequent than *well* in all three main subsections of MEMT, although the differences in normalized frequencies are small. In the verse texts, *full* has a much higher incidence than the other two amplifiers. It is familiar as a conventional metre filler in medieval poetry, and this explains its frequency in medical verse texts as well. The results also reflect the ongoing diachronic shift that was taking place in amplifier use in the ME period (see Section 2.4). The early fourteenth-century material in the appendix favours *well* instead of *full* and *right*, whereas in the later prose materials *well* has lower frequencies than its rivals. The differences are, however, quite small.

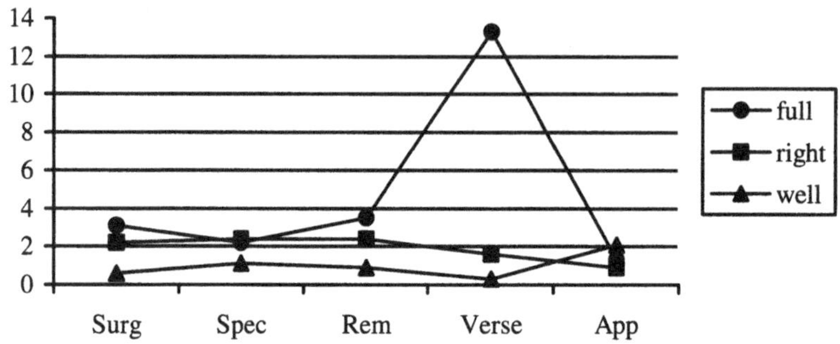

Figure 5. *Full*, *right* and *well* in subsections of MEMT; normalized figures per 10,000 words.

6. Discourse contexts and domain-specific use

The analysis of amplifiers in their microlevel context provides information about their domain-specific communicative functions in medieval medical discourse. As the material in MEMT cuts across the whole domain, containing texts from different subfields of medieval medicine and their traditions of writing, and texts with different communicative roles ranging from expository to instructive, there is potential for great variation in the functions of amplifiers.

For this study, I have examined the discursive uses of three amplifiers: *full*, *right* and *well*. The analysis shows that they occur in similar contexts and with largely similar functional distributions across the material. Three partly overlapping communicative contexts are particularly prominent. In these contexts, amplifiers are used for scaling up various descriptive, instructive and persuasive domain-specific meanings, illustrated in the following subsections.

6.1. Descriptive contexts

Descriptions and categorizations are central in the organization of knowledge in any field of science. In medicine, where science and craft meet, precise descriptions and clear distinctions are also crucial for successful medical practice. In MEMT material, amplifiers frequently occur in contexts providing descriptive information or underlining distinctions. They can be used for highlighting particular characteristics of, for example, body parts, physiological processes or sicknesses, or for underlining specific qualities of substances and instruments. Examples can be found in texts from all traditions of medical writing.

(16) Of which þe first of ossis sacri is *riȝt* grosse, & þo þat foloweþ toward þe lure & þe ende of þe backe (Chauliac, *Anatomy 1*, p. 132)

(17) The skyn is a membre consimile or official, compouned, purely spermatik, strong & towȝ, mene bitwene hardnesse and neischenesse, flexible & *riȝt* sensible; þinne & *riȝt* temperat in complexioun (*Chirurgie de 1392*, p. 12b)

(18) A fe[ver] peracut is a *wele* scharp fe[ver], id est a fe[ver] *wondyr* scharp & wode (Daniel, *Liber uricrisiarum 1*, p. 142)

(19) the propirte of rede coler, the body forsoth wherof it is compownd and made is white and rede to the colour citryne, the heris like citryne, gret veynes and large, and *ful* rede eyen (*De spermate*, p. 195)

(20) Letuse ys colde and *right* moist (Macer, *De viribus herbarum*, p. 90)

Sometimes an amplifier scales up descriptions of signs and symptoms of ailments, also crucial in medical practice.

(21) þer appered aboue ȝe emplastre *ful* putrid quitour in superflue quantite. (Arderne, *Fistula*, p. 34)

(22) And his yȝgen watren and ben *riȝt* soor and he haþ an hote swete in þe delyuerance of his axes. (Gilbertus Anglicus, *Compendium medicinae*, p. 22)

(23) And þerfore it makeþ a mon *ful* hot wyþ-outen-forþ, & to hafe a swyft breþyng (Gilbertus Anglicus, *Compendium medicinae* (*epilepsy*), p. 237)

In some cases an amplifier is used to underline the difficulties or dangers that can be encountered in treatment. Only *full* occurs in this function in the data.

(24) þai holden ham bounden wiþ cordes imade large, whos kyttynge is *ful* harmeful (Chauliac, *Cyrurgie*, p. 52)

(25) hit is *ful* harde or ellis inpossible to be curid (Rupescissa, *Remedies*, p. 122)

6.2. Instructive contexts

Amplifiers are frequent in instructive passages across the material. These passages most often give advice on how to prepare medicines or how to treat patients, administer medicines or perform operations. Some overlap with descriptive contexts is evident in the examples.

(26) brek al-togidre in a brasen morter into *ful* smal poudre (Arderne, *Clysters*, p. 100)

(27) temper þem with the whitte of egges tyll þei be *wele* liquyde. (Benvenutus Grassus, *De probatissima arte oculorum*, p. 77)

(28) stampe thyne herbes and þe mirr *riȝt* small & lete sethe hem a while (*Sekenesse of Wymmen 3*, p. 72)

(29) & wan yt ys *well* cold anoynte þ[e] sore þer with (*Killeen Medical Texts*, p. 156)

(30) let it kole and after it is *right* colde stire in it (*Leechbook 1*, p. 34)

Instructive contexts also include warnings and admonitions for practitioners. The phrase *well ware*, an early variant of the PDE fixed collocation *well aware*, is common, as examples (31)-(33) show. The only instance of the adjective *ware* modified by another amplifier (34) occurs in the language of William Caxton, who is well known for his old-fashioned and sometimes peculiar language use.

(31) The secunde rule: be *wel* war inas-miche as þou maist, þat þer be no seem ne no foldynge, neiþir silfegge in þe rolle þat mowe greue þe pacient. (Mondeville, *Chirurgie*, p. 282a)

(32) And be *wele* ware er thow ete eft sones, that thy stomak be clene and voide fro the first mete (*Secrete of Secretes*, p. 55)

(33) and be *weel* war þat non yren touche þe watir (*Quinte Essence*, p. 5)

(34) and desyre wyth thy helpe / to be *ryght* ware herafter that I fall not in to syn (Caxton, *Ars moriendi*)

6.3. Persuasive contexts

Persuasive uses of amplifiers in the material are often connected with evaluations. Amplifiers frequently co-occur with evaluative adjectives like *necessary*, *good* or *profitable* to emphasize the importance or usefulness of particular measures, treatments or substances for the patient's health, for a specific group of patients, or for curing particular troubles. Examples (35)-(39) illustrate amplifiers in this function in different traditions of writing.

(35) if þe pacient may abstene hymself fro þe pryué by two dayes, it war *full* necessary to hym. (Arderne, *Fistula*, p. 35)

(36) And he calleþ yt Alabaustrum. And it is *riȝt* propre to contraxions of nerueȝ. (Chauliac, *Wounds*, p. 44)

(37) And not oonly of precyous stonys but also of gommys and of oþer thyngys may be made *ful* holsom powders for eyon. (Benvenutus Grassus, *De probatissima arte oculorum*, p. 88)

(38) The thirde is *well* profytable and *wel* nedeful that all that shall be etyn be wele and smale chewed (Caxton, *Gouernayle of Helthe*, f. A1)

(39) for thys hath provyd *ryȝt* good dyuerse tymys in dyuerse personys. (*Wyse Boke of Maystyr Peers of Salerne*, p. 314)

Amplifiers can also co-occur with evaluative adjectives in contexts referring to authoritative sources of knowledge.

(40) Oure lord Ihesu y-blessid God knoweth that I lye not, and therfore no man dout of this, þof-al old famous men and *ful* clere in studie haue confessed tham that thei fande nat the wey of curacion in this case. (Arderne, *Fistula*, p. 3)

(41) Of them the other philosophres that have be sithe toke thair begynnyng, bothe Indiens, Persiens, Grekes, and Latynes, and wrote vpon thair principles, of artis and sciences the secretes, for in their writyng is no thynge founed, no thynge reproved, but of *full* wise men approved. (*Secrete of Secretes*, p. 48)

(42) Fulgens discryuyth exercise in a *full* fayre maner saying in this wise. (*Gouernayle of Helthe*, f. 12v)

7. Conclusion

In this study, I have examined the use of amplifiers in *Middle English Medical Texts* (MEMT), an electronic corpus of late medieval medical writing. Combining contextualized quantitative and qualitative analyses, my aim has been to provide an overview of amplifiers in the material, and to highlight some patterns of variation in the use of individual items and between different traditions of medical writing. I have focused on amplifiers in adjective modification, which is their prototypical function in PDE.

From a pre-defined search list of 57 items, collected from earlier studies, 28 amplifiers occur in MEMT with an adjectival head. The widest repertoire is attested in the category of remedies and *materia medica*, while texts derived from learned traditions, i.e. surgical and other specialized treatises, have a more restricted range. Verse texts have fewer types than prose texts. On the other hand, the relative frequency of amplifiers in verse texts is almost twice as high as in prose texts, largely due to their use as metre fillers. In the prose traditions, amplifiers occur with a slightly higher relative frequency in remedy texts than in surgical and specialized texts.

The frequencies of individual amplifiers vary greatly. The most frequent item is *full*, which according to earlier studies on other types of texts is the predominant adjectival amplifier in this period. *Full* covers almost 30% of all amplifier data in MEMT. In the frequency ranking it is followed by *right* (17.4%) and *so* (14%), then *all* (11.8%) and *well* (6.9%). Most amplifiers have low frequencies, occurring less than 20 times in the half-a-million-word corpus. For example, *very*, the predominant booster from the sixteenth century onwards, is still rare. Many amplifiers occur just once.

An analysis of the discursive uses of three amplifiers, *full*, *right* and *well*, reveals three partly overlapping communicative contexts where amplifiers tend to occur in MEMT. Firstly, amplifiers are common in descriptive and expository contexts, foregrounding specific characteristics of relevant objects and concepts, and distinctions between them. Secondly, they are frequent in instructive passages, highlighting advice on medical praxis. Thirdly, amplifiers are commonly used for rhetorical purposes. They frequently co-occur with evaluative adjectives in contexts emphasizing the importance or usefulness of particular medical treatments or substances.

In a diachronic perspective, some interesting comparisons can be made in amplifier use between medieval medical texts in MEMT and present-day medical texts in Medicor, a corpus of late twentieth-century American English medical writing (see Pahta forthcoming b). The overall frequency of amplifiers occurring with adjectival heads in the two corpora is surprisingly similar: MEMT contains 12.7 amplifiers for every 10,000 words, while the corresponding figure in Medicor is 12.4. In both datasets, amplifiers in general tend to be more frequent in texts that can be characterized as more 'popular' rather than 'academic', although there is great variation from one text to another. Furthermore, amplifiers tend to occur in similar kinds of discursive contexts and communicative functions in the two datasets, underlining descriptions and distinctions, emphasizing instructions and advice, and persuading readers of the value of particular methods or measures by scaling up evaluations. It is also interesting to note that in both corpora the repertoire of amplifiers is wide, whereas in terms of frequency, a few items in each period dominate. The items leading the frequency ranking in the two periods are different. The most frequent medieval amplifier *full(y)* is rare in PDE medical texts and *right* does not occur at all, while the two most common present-day amplifiers, *very* and *highly*, are not frequent in the medieval data. The amplifier *well*, on the other hand, ranks high in both periods, and another interesting item with a relatively high frequency in both datasets is *so*. Future studies will throw more light on the use of these items, as well as on other developments in amplifier use in the special language of medicine across time.

References

Aarts, Bas / Denison, David / Keizer, Evelien / Popova, Gergana (eds) 2004. *Fuzzy Grammar: A Reader*. Oxford: Oxford Univ. Press.

Allerton, David J. 1987. English Intensifiers and their Idiosyncrasies. In Steele, Ross / Threadgold, Terry (eds) *Language Topics: Essays in Honour of Michael Halliday*. Amsterdam: Benjamins, 2/15-31.

Bäcklund, Ulf 1973. *The Collocation of Adverbs of Degree in English*. Uppsala: Almqvist & Wiksell.

Biber, Douglas / Johansson, Stig / Leech, Geoffrey / Conrad, Susan / Finegan, Edward 1999. *Longman Grammar of Spoken and Written English*. London: Longman.

Bolinger, Dwight 1972. *Degree Words*. The Hague: Mouton.

Chafe, Wallace 1987. Evidentiality in English Conversation and Academic Writing. In Chafe, Wallace / Nicholls, Johanna (eds) *Evidentiality: The Linguistic Coding of Epistemology*. Norwood, NJ: Ablex, 261-272.

Gonzáles-Díaz, Victorina / Adamson, Sylvia 2004. Back to the *Very* Beginning: The Development of Intensifiers in Middle English. Paper presented at the 13th International Conference of English Historical Linguistics, Vienna, 23-28 August.

Hopper, Paul J. / Traugott, Elizabeth C. ^{2}2003. *Grammaticalization*. Cambridge: Cambridge University Press.

Hyland, Ken 2000. *Disciplinary Discourses: Social Interactions in Academic Writing*. Harlow: Pearson Education.

Ito, Rika / Tagliamonte, Sali 2003. *Well* weird, *right* dodgy, *very* strange, *really* cool: Layering and Recycling in English Intensifiers. *Language in Society* 32, 257-279.

Labov, William 1984. Intensity. In Schiffrin, Deborah (ed.) *Meaning, Form and Use in Context: Linguistic Applications*. Washington, DC: Georgetown University Press, 43-70.

Lorenz, Gunter R. 1999. *Adjective Intensification – Learners versus Native Speakers: A Corpus Study of Argumentative Writing*. Amsterdam: Rodopi.

Lorenz, Gunter R. 2002. *Really Worthwhile* or *Not Really Significant*? A Corpus-based Approach to the Delexicalization and Grammaticalization of Intensifiers in Modern English. In Wischer, Ilse / Diewald, Gabriele (eds) *New Reflections on Grammaticalization*. Amsterdam: Benjamins, 143-161.

Méndez-Naya, Belén 2003. On Intensifiers and Grammaticalization: The Case of *swiþe*. *English Studies* 84, 372-391.

Méndez-Naya, Belén 2004. *Full Good, Right Good, Well Good*? On the Competition of Intensifiers in the Middle English Period. Paper presented at the 13th International Conference of English Historical Linguistics, Vienna, 23-28 August

Méndez-Naya, Belén 2005. *He Was Nat Right Fat*: On the Origin and Development of Intensifier *Right*. Paper presented at the 5th International Conference on Middle English, Naples, 23-26 August.

Mustanoja, T.F. 1960. *A Middle English Syntax*. Helsinki: Société Néophilologique.

Nevalainen, Terttu 1997. The Processes of Adverb Derivation in Late Middle English and Early Modern English. In Rissanen, Matti / Kytö, Merja / Heikkonen, Kirsi (eds) 1997. *Grammaticalization at Work: Studies of Long-term Developments in English*. Berlin: Mouton, 145-190.

Norri, Juhani 1992. *Names of Sicknesses in English, 1400-1550: An Exploration of the Lexical Field*. Helsinki: Academia Scientiarum Fennica.

Pahta, Päivi forthcoming a. Amplifiers in Early Modern English Medical Texts. In Taavitsainen, Irma / Pahta, Päivi (eds) *Argumentation in Early Modern English Medical Writing*.

Pahta, Päivi forthcoming b. *This is Very Important*: A Corpus Study of Amplifiers in Medical Writing. In Gotti, Maurizio / Salager-Meyer, Françoise (eds) *Advances in Medical Discourse Analysis: Oral and Written Contexts*. Bern: Peter Lang.

Pahta, Päivi / Taavitsainen, Irma 2004. Vernacularization of Scientific and Medical Writing in its Sociohistorical Context. In Taavitsainen, Irma / Pahta, Päivi (eds), *Medical and Scientific Writing in Late Medieval English*. Cambridge: Cambridge University Press, 1-18.

Paradis, Carita 1997. *Degree Modifiers of Adjectives in Spoken British English*. Lund: Lund University Press.

Partington, Alan 1993. Corpus Evidence of Language Change: The Case of Intensifiers. In Baker, Mona / Francis, Gill / Tognini-Bonelli, Elena (eds) *Text and Technology: In Honour of John Sinclair*. Amsterdam: Benjamins, 172-199.

Peters, Hans 1994. Degree Adverbs in Early Modern English. In Kastovsky, Dieter (ed.) *Studies in Early Modern English*. Berlin: Mouton, 269-288.

Quirk, Randolph / Greenbaum, Sidney / Leech, Geoffrey / Svartvik, Jan 1985. *A Comprehensive Grammar of the English Language*. London: Longman.

Rissanen, Matti / Kytö, Merja / Palander-Collin, Minna (eds) 1993. *Early English in the Computer Age: Explorations through the Helsinki Corpus*. Berlin: Mouton.

Taavitsainen, Irma / Pahta, Päivi 1998. Vernacularization of Medical Writing in English: A Corpus-Based Study of Scholasticism. *Early Science and Medicine* 3, 157-185.

Taavitsainen, Irma / Pahta, Päivi / Mäkinen, Martti 2005. *Middle English Medical Texts*. CD-ROM with MEMT Presenter by Raymond Hickey. Amsterdam: Benjamins.

Taavitsainen, Irma / Pahta, Päivi / Mäkinen, Martti 2006. Towards a Corpus-based History of Specialized Languages: *Middle English Medical Texts*. In Facchinetti, Roberta / Rissanen, Matti (eds) *Corpus-based Studies in Diachronic English*. Bern: Peter Lang, 79-93.

Tagliamonte, Sali / Roberts, Chris 2005. *So* Weird; *So* Cool; *So* Innovative: The Use of Intensifiers in the Television Series *Friends*. *American Speech* 80, 280-300.

Voigts, Linda Ehrsam 1984. Medical Prose. In Edwards, A.S.G. (ed.) *Middle English Prose: A Critical Guide to Major Authors and Genres*. New Brunswick, NJ: Rutgers Univ. Press, 315-335.

Voigts, Linda Ehrsam 1989. Scientific and Medical Books. In Griffiths, Jeremy / Pearsall, Derek (eds) *Book Production and Publishing in Britain, 1375-1475*. Cambridge: Cambridge Univ. Press, 345-402.

JUKKA TYRKKÖ

From *Tokens* to *Symptoms*: 300 Years of Developing Discourse on Medical Diagnosis in English Medical Writing

1. Introduction

Medicine was the first branch of learning to focus theoretical attention on the concept of *sign* more than two thousand years ago. Ever since that time, the study of signs has been of fundamental importance to the practice of medicine. This long, unbroken presence of semiotic[1] issues in medical writing makes signification a particularly fruitful area for diachronic studies of the development of specialist medical language; concurrently, it allows insights into the development of the medical community from loose and heterogenous groups of practitioners into a true professional community marked by a firmly established specialist lexicon.

This chapter examines the history of medical signifier terms using a corpus linguistic approach. I will begin by briefly surveying the historical context and theoretical landscape of medical semiotics. Next, through data derived from the Corpus of Early English Medical Writing (CEEM), I will discuss the most important signifiers *token*, *sign*, *symptom*, and *accident* in vernacular English medical writing from the end of the fourteenth century to the beginning of the eighteenth century with the objective of outlining their actual usage. The findings will be related to developments in both medical science and the professional medical community, focusing on the influence that each had on linguistic practice and semantic changes. Signifiers

1 Modern semiotics derives its primary intellectual heritage from philosophy and linguistics. Sebeok (1976: 117-148) defines medicine as the third part of an encompassing theory of signs, along with philosophy and linguistics.

are in *italics* when referred to as lexical items, in normal typeface when discussed as semiotic concepts.

2. Medical semiotics

The field of medical semiotics can be given three definitions. In the broadest sense, medical semiotics refers to the study of signification in medical science; particularly, semiotic fields of medicine include symptomology, the study of symptoms – and nosology, the study of disease classification. Somewhat more particularly, medical semiotics can be defined as the study of signifier terminology and the way such terms are employed in the description of diseases and injuries. Thirdly, according to the narrowest definition, medical semiotics refers to a specific branch of medicine which enjoyed a fair bit of popularity from the late seventeenth century to the early nineteenth century, particularly in France, Germany and the Netherlands and to a somewhat lesser extent in Britain.[2] This chapter discusses medical semiotics in the intermediate sense, focusing on the signifiers used in medical diagnosis and prognosis.

The successful practice of medicine relies in part on the correct interpretation of *signs* observed by the doctor and on information gathered in a medical interview during which the patient is asked about his or her previous medical history (*anamnesis*), circumstances preceding the current ailment and the *symptoms* experienced. The conceptual origins of both sign and symptom can be traced back to classical Greece: the lexical items σημείο (semeio) and σύμπτωμα (symptoma) are found in medical texts from the *Corpus Hippocraticum* onwards.[3] The terms were latinized as *signum* and *symptoma*,[4]

2 Rather than focusing on diseases and their description, semiotic medicine was preoccupied with listing and classifying medical signs and symptoms. On the history of medical semiotics, see Eich (1986), Hess (1993 and 1998), Sebeok (2001) and Vila (2003).

3 Σημείο appears in the *Corpus Hippocraticum* 260 times, σύμπτωμα only twice (Maloney / Frohn 1986-1989). The Greek lexicon had many other signifier

respectively. The two were essentially synonymous, a symptom being a sign which requires interpretation by the doctor (Sebeok 1999: 50-54). By making a theoretical distinction not only between the observed and what it stands for, but also between indexical signs and signification requiring knowledge-based interpretation, Hippocratic physicians laid the foundation for the concept of disease as something separate from the immediate complaint. It is noteworthy that these classical senses of sign and symptom are quite close to their modern meanings. Modern medicine defines signs and symptoms as manifestations of physiological processes, which in turn are the causes of illnesses. *Sign* denotes an objective clinical or physical feature observed by the doctor, *symptom* a subjective one reported by the patient.[5] In view of this apparent semantic continuity, it would seem reasonable to say that the core terminology of medical semiotics has survived relatively unchanged for 2500 years. That, however, is not quite the whole story; in actual fact, the lexical items have seen a number of changes of meaning along the way.

Medieval medical practice in England was based on the classical humoral model, albeit in an abbreviated and somewhat corrupted form.[6] The concepts of disease and sign were not differentiated from each other as they are today (Norri 1992: 87; Jones

terms as well (such as ειδοσ, form) but the others did not affect early English (cf. Sebeok 2001: 52).

4 *Symptom* appears transiently as *sinthoma* in medieval Latin. Nutton (1993: 7) suggests the translation 'happening' for *symptom*, i.e. something which happens during or in connection with a disease. *Accident*, the medieval translation of *symptom*, derives from this meaning (see below).

5 These definitions, found both in the *Oxford English Dictionary* (2nd edition, online; henceforth OED) and current medical dictionaries and textbooks, are universally accepted in modern medical lexicon. Whether this means that the distinction is always observed in practical usage of these terms is of course another matter; cf. Staiano (1982: 332), Rausch (1999: 111).

6 While some medical texts survived in monasteries, the stock of texts was fragmented and severely lacking in comparison to the wealth of original Greek medical writing. New texts contributing to a more comprehensive understanding of classical medicine started becoming available from mainland Europe toward the end of the medieval period, but for the most part the theoretical system taught even in the universities was narrow.

1984: 58). The role of Galen,[7] the uncontested medical authority for well over a thousand years, is crucial in explaining the medieval and Renaissance understanding of symptom. Galen advocated the view that everything out of the ordinary in the human body should be seen as a sign and that signs in the medical context should be called *symptoma* (Hess 1993: 54; Sebeok 1999: 53).[8] Even more importantly, he also argued that actual instances of 'harm' to bodily functions should be defined as diseases and not merely as signs (Galenus 2004: 228). The resulting confusion in terminology transformed the previously synonymous relationship of symptom and sign: symptom came to stand for both a sign and for a disease which co-occurs with another.[9]

This model prevailed until the early seventeenth century, at which time the empirical movement started changing medicine by placing increased emphasis on experimentation and direct observation. New, evidence-based knowledge was increasingly accepted by medical authorities and Galen's infallibility could be openly questioned. By focusing on first-hand observation, empiricist also raised interest in semiotics: instead of simply learning the signs and symptoms of each disease by heart, physicians were encouraged to study and identify the signs of diseases for themselves. In the process, *symptom*'s meaning as a co-disease which could not be explained through humoral theory, became outdated.[10] Around 1750, the latest

7 Galen of Pergamon (131-201 A.D.) was born in Greece but made a career for himself in Rome. His influence lasted virtually uncontested until the end of the seventeenth century. An admirer of Hippocrates, Galen added philosophical aspects to Hippocratic theory and contributed to all branches of medicine from anatomy and neurology to humoral theory (cf. French 2003: 34-58; Siraisi 1990).

8 Galen divided medical practice into six branches, one of which was semiotics. He further classified semiotics into the three categories of diagnosis, anamnesis, and prognosis (Sebeok 1999: 53; cf. Foucault 1973: 90).

9 The closest modern equivalent to this latter definition of a symptom is *complication*, generally defined as a condition resulting from a previous disease or injury, sometimes as a consequence of treatment. By contrast, the classical co-occurring affect was a simultaneous complaint.

10 The discoveries of empiricists, particularly in the field of contagious diseases, finally led to the end of humoral medicine (French 2003: 158-159), although this happened very gradually and humoral theory only truly succumbed in the

date covered by this study, *symptom* had once again become more or less equivalent with *sign*, as it had been in antiquity.[11]

3. Material and methodology

The data for the present study comes from an electronic corpus of ME and EModE medical texts. The *Corpus of Early English Medical Writing* (CEEM) is a purpose-built corpus of English vernacular medical texts from c. 1375 to c. 1750. The first part, *Middle English Medical Texts* (MEMT), was published on CD-Rom in late 2005, while the second part on EModE is currently being compiled by the Scientific Thought Styles project at the Research unit for Variation and Change in English at the University of Helsinki.[12] The current size of the full corpus is 1.8 million words, representing medical writing from remedy texts to learned academic treatises on specialized topics and on surgery. The 350-year time span is divided into four periods and the texts into three traditions of transmission: specialized academic treatises, surgical and remedy texts. This division was introduced by Voigts (1982 and 1984) and has become widely adopted in the discussion of early vernacular medical writing (Norri 1992: 34-36; Taavitsainen / Pahta 2004: 14-15; see also Mäkinen, this volume). As the compilation of the Early Modern part of the corpus (1660-1750) is in progress, although it is already quite representative up to

late nineteenth century following Pasteur's discovery of microbes (Thagard 1999: 24-26).

11 Later, towards the end of the eighteenth century, many physicians understood *symptoms* as manifest signifiers and *signs* as the interpretations drawn on the basis of them by the doctors (Risse 1987: 116). The early nineteenth century saw the last semantic development as a result of the growing role of medical signs produced by laboratory medicine: *sign* came to denote objective scientifically verified indications and *symptom* began to be used for indications reported by the patient.

12 The project is led by Irma Taavitsainen and Päivi Pahta; current project members are Martti Mäkinen, Carla Suhr, Maura Ratia, Jukka Tyrkkö and Turo Hiltunen.

1700, results from this period must be considered tentative rather than conclusive.

The present study commenced with the compilation of a unique word list, a process which yields a list of every spelling variation and inflected form of every lemma in the corpus as a unique entry. The high frequency of the Romanic loan *accident* was discovered at this stage and as a result the term was included in the study.[13] The numerous spelling variations encountered in historical material (especially from the ME period) rule out the possibility of simply searching for previously known variants or even the use of conventional plural endings. Once the list was available, relevant word forms were compiled into search word lists which were used as the basis of the actual lexical searches. The results were manually verified, contextually false items were discarded and the remaining occurrences were then collected in spreadsheets for analysis. Frequencies (see Appendix – Figure 1) were calculated for both the entire material and for individual time periods and textual traditions.[14]

The occasional remarks on the occurrences and definitions of the signifier terms in early dictionaries and glossaries are based on my earlier study of signifier terms in dictionaries.[15]

---

13 Although *accident* was already known to have had a signifier meaning in early medical writing, the frequency came as a surprise because the term is very rarely mentioned in discussions of medical language. This finding is an outstanding example of how corpus material can reveal quantitative information which might otherwise be overlooked.

14 Frequency counts have limited reliability as a means of studying relatively rare lexical items in an extract corpus. The nature of the text and the part from which each individual example was extracted dictate the frequency of such items: for example, a prologue would be expected to contain considerably fewer instances of names of surgical instruments than a description of a particular surgical procedure found later in the text. For frequencies in the different textual traditions, see Section 5.

15 For more on the representation of the early meanings of signifier terms in the OED and in early dictionaries from Palsgrave (1530) to Scott-Bailey (1772), see Tyrkkö (2006).

4. Signification in English medical discourse

For the first two hundred years of scientific vernacularization which began at the end of the fourteenth century, learned medical texts were rarely original compositions. Instead, they were derivative compositions, extractions, and commentaries of translated Latin and French texts (cf. Taavitsainen 2004). As a general rule, a medical treatise would begin with a description of an ailment, followed by the treatment and a section on prognosis. Signs and symptoms are typically mentioned in both the diagnostic and the prognostic part of a text; in the former they are descriptive features, in the latter they signal the likely outcome of the disease process, particularly the likelihood of life or death. Prognostic sections were frequently characterized by the postmodification of signifier terms with lexical items like 'evil' or 'death'. Diseases were often described with formulaic syntactic structures, as in (1) from Gilbertus Anglicus (fifteenth century). The formulaic structure (in bold) is repeated in the text for each disease described:

(1) But if it is of colere, **þese ben þe tokenes**: grete wille to fiȝte and to smyte, drines of þe mouþe, blacknes of þe tonge, myche sharpe and bitter ache, and moche stering of þe hert.

Naturally, semiotic signification does not necessarily require the explicit use of signifier terms. This is important particularly in light of the ostensible differences in the overall frequencies of signifier terms in medical texts at various periods. A lower frequency of signifiers does not automatically mean that diagnostic or semiotic issues were not discussed, but rather that the discussion may have been conducted through other discoursive means, e.g. verbs of knowledge, as in (2):

(2) & þou schalt **knowe** a cold dicrasie **bi** þe whiȝtnes eiþer þe wannesse of þe place & whanne þe place is cold in towchinge, & whanne þat þe quytture is þicke; þe which discrasie þou schalt remeue bi þe enoynture of hote oiles

Fifteenth- and sixteenth-century scientific writing was characterized by the abundance of medical neologisms (Norri 1992: 24-30; McConchie 1997: 14-61). The need for new terminology was considerable, because Latin texts contained concepts for which no vernacular equivalent existed. Vernacular texts were not only intended for readers who were unable to read Latin, but for a diverse readership ranging from university educated doctors to unlearned healers and even regular householders. The heterogeneity of audiences led translators and writers to employ a rich variety in lexical choices, some preferring items from the existing vernacular lexicon, others opting for Latin or French loans for reasons of precision or prestige (cf. Jackson 2002: 39) – early medical writing shows evidence of both strategies. Like all languages, ME already had a stock of native signifier terms and thus borrowing would not have been necessary to fill a complete *lacuna*: the OE lexical item *token* featured prominently in vernacular medical writing as a translation of *sign* all the way up to the mid-sixteenth century. The case is somewhat different when it comes to the Romanic loan *accident*, which was used as a translation of *symptom* from the very end of the fourteenth century. Unlike *sign*, *symptom* revealed a genuine conceptual gap in the vernacular. While the primary reason for using *accident* was to fill this lexical need, a secondary motive may have been to preserve some of the prestige value of a foreign term by translating it with another, more transparent one. The entry of the four terms into English stretches over a considerable time span: *token* is an Anglo-Saxon gloss and dates back to the late ninth century; *sign* is first recorded in 1225; *accident* and *symptom* are both first recorded during the last quarter of the fourteenth century, thus appearing in English when vernacular scientific writing begins (Tyrkkö 2006). An initial statistical analysis of the corpus material indicated that these lexical items account for the overwhelming majority of signifier terms in the corpus and thus represent the key signifier terminology used by authors and translators of medical texts between 1375 and 1725.

4.1. Signifiers before 1550

The vernacularization of scientific texts, a process in which medical writing was at the forefront, began around 1375 and came to an important threshold at the end of the fifteenth century with the advent of the printing press and subsequent wider availability of texts. By

1550, English had become accepted as the language of original scientific compositions and the medieval way of thinking in medicine had started to give way to empiricism and an inductive thought style (Taavitsainen / Pahta 1998: 165). Most of the vernacular medical texts published in print between 1475 and 1550 were still based on manuscripts circulated during the medieval period (Norri 1992: 54). Academic treatises intended for international audiences were written almost exclusively in Latin, but English soon became a valid language for serious scientific writing (McConchie 1997: 14-23, 65-80) and the number of translations grew rapidly. The first bilingual dictionaries (such as the *Promptorium Parvulorum* and *Hortus Vocabulorum*) date back to the middle of this period but information on word meanings can also be found in small word lists and glossaries occasionally published as appendices to larger works, some of which were medical in nature.[16]

The frequency counts (see Appendix – Figure 1) show that *token* and *accident* were preferred in vernacular writing over *sign* and *symptom* during the late ME period. Considering the primary reason behind translations – to make texts available to an audience unfamiliar with a foreign language or simply foreign terminology – the fact that the Anglo-Saxon *token* was more frequent than the Latinate *sign* is perhaps not too surprising, but it does suggest that *sign* had not been fully assimilated into the vernacular lexicon.[17] The corpus contains 62 texts dated between 1375 and 1550; 30 of them include *token*, 24 include *sign*, and 17 texts include both *token* and *sign*. Bearing in mind that these figures represent exclusively medical instances of the lexical items, the fact that many texts would make use of both *token* and *sign* shows the capricious nature of early vernacular writing. With *accident* and *symptom* things get a little more complicated. To begin with, the data shows unequivocally that *accident*, a French loan[18]

16 E.g. the glossary compiled by Bartholomew Traheron for de Vigo's *The Most Excellent Workes of Chirurgerye* (1543; see References, primary source 5).

17 The OED dates the entry of *sign* to 1225, thus giving it a 150-year history in the vernacular prior to the writing of the earliest texts in CEEM. Interestingly, the corpus does not contain a single incidence of a corresponding Latin code-switch (*signum* or *signa*).

18 The OED claims *accident* is a French loan. This may be so, but Latin and Italian are other possible sources, particularly considering the importance of

dating back to 1374 (OED), was the established term for the concurrent medical ailment and sign between 1375 and 1550. In light of the fact that the OED dates the first medical usage of *accident* to 1563, we may say that the evidence draws attention to a previously unrecorded piece of lexical history. By contrast, the data shows surprising scarcity of *accident*'s classical predecessor *symptom*, which occurs only 10 times in 3 texts between 1375 and 1475 and not once thereafter until the end of the sixteenth century. All 10 occurrences follow the Medieval Latin spelling *sinthoma or sinthomate*, the uninflected form suggesting a code-switch rather than a borrowing.[19] Trevisa's *On the properties of things*[20] (1398) is the earliest known English text to include *symptom*:

(3) For þe heed is more and roote of al þe body and first and principal foundement of al þe bodiliche vertues; ȝif þe heed is wel disposed and i-ordeyned, al þat ben þervndir ben in þe bettir disposicioun and state. And aȝenward, ȝif þe heed is corrupt and distempered with **sinthoma** of corrupcioun and hedeache, nedis þe neþer membres of þe body ben desesed.

Considering the wealth of occurrences for *accident* over the same time period, there can be no doubt that as far as *symptom* is concerned, translation rather than borrowing was the overwhelmingly preferred practice. While over-ambitious generalizations must always be

---

the universities of Salerno and Bologna as centers of medieval medical learning, especially in the field of surgery (Pahta 1998: 12, 23). At the very least the medical meaning of *accident* is likely to have been influenced by Italian terminology.

19 Code-switches in early scientific writing can be motivated by three main reasons: jargon, quotations and intertextuality, and textual organization (Pahta 2004). The occurrences of Latin *sinthoma* show features of both jargon and intertextuality, the latter particularly in the occurrences in Trevisa (1398). The OED records the medieval spelling for all occurrences of *symptom* prior to 1541 and for several thereafter. The last citation of the intermediate form *symptoma* in the OED is Daniel's *The queenes arcadia* (1605). On *sinthoma* see also Norri (1990: 118).

20 It is worth noting this is not an exclusively medical text. While it includes sections on medicine, it is an encyclopedic work and represented the highest strata of knowledge at its time. A translation of Bartholomeus Anglicus' *De proprietatibus rerum* (c 1240), Trevisa's text is the source of 820 citations in the OED.

guarded against when drawing conclusions on the basis of corpora, it is safe to say that *symptom* was a rare term in medical writing up until the very end of the sixteenth century. *Symptom* can be found in only one early dictionary or glossary prior to 1550, Elyot's Latin-English *Dictionary of Syr Thomas Elyot* (1538) – *accident*, by contrast, can be found in several glossaries. This suggests that *symptom* was so rare in texts that early lexicographers either simply didn't encounter it or else excluded it on the grounds that it was a foreign word.[21]

4.2. Signifiers between 1550 and 1750

The conventions of vernacular medical writing reflected the increasing role of the medical community during the late sixteenth century. While writers still favored Latin in texts aimed for the international medical community, English was now widely accepted as a valid language for serious medical writing (French 2003: 204). The growing number of printed books and the wider availability of texts meant that a professional community, marked by a specialist lexicon, was being established even more firmly than before. The budding empirical movement brought along a re-evaluation of previously untouchable authorities and shifted the focus on personal observations rather than reverence for established beliefs. Comprehensive dictionaries, both general and specialist, also began to appear during this period and thus for the first time we have a source of explicit definitions to which we can turn when discussing the contemporary meanings of specialized terminology.

The findings from this period illustrate the extent to which the middle of the sixteenth century was a linguistic threshold in medical writing. The frequency of *token* went into a steep decline around 1550, indicating the pace at which medical writing was becoming averse to lexical items from the vernacular lexicon. *Sign*, by contrast,

---

21 Most early glossaries and dictionaries were compiled for the practical purpose of either helping readers decipher hard words or to aid language learning (Landau 2001: 45-46). Lexicographical practices varied greatly when it came to foreign words: some lexicographers included them, others more concerned with lexical purity very deliberately left them out (Jackson 2002: 33-42).

was the most frequent signifier during the 1550-1660 period. A similarly straightforward process of semantic change cannot be observed with *accident* and *symptom*. While the reappearance of *symptom* after 1550 can be construed as testament to a more markedly professional lexicon and the broader availability of newly translated classical texts, the frequency of *accident* also increased for the 1550-1660 period and only started to decline thereafter. Unlike *symptom*, *accident* did not make a semantic shift from co-occurring ailment to sign, but instead dropped its medical meaning during the early eighteenth century and settled on its modern meaning. In dictionaries and glossaries, *accident* is recorded more frequently than *symptom* prior to Cockeram's *English Dictionary* (1623) and less frequently thereafter, indicating that *accident* had become a common word while *symptom* was only beginning to appear in English texts and thus needed definitions. *Accident* is generally defined as 'chance'; the medical meaning is generally included only in specifically medical glossaries. *Symptom* appears for the first time in a monolingual dictionary in Cawdrey's *A Table Alphabeticall* (1604) and both *symptom* and *accident* can be found in the major monolingual dictionaries from Cawdrey (1604) onward. The equivalence of the two terms is evident from the definitions given in contemporary medical dictionaries: most definitions for *symptom* make use of the term *accident*, reinforcing the fact that *accident* was considered the more familiar and established term. By contrast, *accident* is not defined as a *symptom* in any early dictionary, indicating that lexicographers probably did not envisage a situation where someone familiar with *symptom* would not know *accident*.

5. Signifiers in medical jargon

Specialized terminology is one of the primary linguistic indications of an established professional community (Garber 2001: 97-148; Swales 1990). In medicine and other sciences, lexical choices reflect not only the need for specialized vocabulary, but also what sources of terminology are available and valued by members of the professional community (Jones 2004: 30-31). The overall patterns of lexical changes (see Appendix – Figures 2 to 4) reflect the lexical changes associated with the development of professional medical communities.

During the late medieval period, the learned medical community wrote primarily in Latin; writing in the vernacular was predominantly practiced either for political purposes or in order to disseminate classical knowledge for those unable to read Latin. As expected, learned vernacular treatises exhibit a higher incidence of Latinate terminology than unlearned remedy texts, but it is worth noting that *token* is also four times more frequent in specialized academic treatises than it is in remedy texts, illustrating the higher role signification plays in learned writing. The 1475 to 1550 period exhibits a very high frequency of both *token* and *sign* in remedy texts, higher in fact than in the learned traditions. *Symptom* is completely absent from the corpus during this period and *accident* only appears in surgical texts. From 1550 onwards, remedy texts exhibit a low overall incidence of signifiers with some slight reflection of the lexical developments in the learned traditions, in which the more academic terms *symptom* and *accident* become increasingly common. The use of *symptom* was strictly limited to the professional register, showing no occurrences in remedy texts prior to 1660 and only a few isolated instances between 1660 and 1725. The specificity of *symptom* as a marker of cutting-edge academic discourse can be illustrated by consulting the part of the corpus containing some 100 articles published in the Royal Academy's *Philosophical Transactions* between 1665 and 1712. The correspondents and editors appear to have clearly favored *symptom*, which accounts for 72% of all occurrences of signifiers in the series. *Sign* accounts for 18%, *accident* for 9% and *token* for 1% (see Appendix – Figure 3).

One particularly curious difference can be observed between the learned traditions. Recent studies have largely discredited the belief that medieval and renaissance surgery was significantly less learned and rational than medicine proper,[22] but some clear-cut terminological

22 Cf. Siraisi (1990: 153-186). Despite the organization of surgeons into guilds which occurred much earlier, surgical skills were not taught in the curriculum of European universities outside Italy until the fifteenth century (Norri 1992:

differences can be observed in the corpus. The most unexpected finding concerns the more complex and higher prestige items *accident* and *symptom*, the latter of which appears to have been almost non-existent in the specialized treatises tradition up until the 1550 mark. This tradition, which contains both translated academic texts and treatises on particular specialized topics, has been found to be the most learned of textual traditions, yet – as far as the choice of signifier terms is concerned – it appears to have been markedly more vernacular than surgical writing. The French loan *accident* in particular enjoyed noticeable popularity in surgical texts, occurring frequently from 1475 onwards. Later, when both *accident* and *symptom* are found in learned traditions, the preference for the former appears to have been a characteristic of surgical writing[23] while academic treatises show a preference for *symptom*. Textual sources and the fact that academic teaching of surgery was at first available only in Italy and France (Pahta 1998: 23) may turn out to be factors explaining this phenomenon – it seems reasonable to conjecture that the lexicon of the surgical tradition would have adopted elements from the languages of regions where their craft was most highly valued. More research is needed.

6. Signifiers and medical science

Having looked at the proportions in which writers preferred one signifier over another, we can now turn to speculating about the reasons behind their choices. One interesting question involves the terms *token* and *sign* and whether their apparent equivalence is a fact or a mere perception. Plenty of examples from the corpus show that

---

49; Pahta 1998: 23). Much of Greek literature on surgery remained inaccessible to Western European surgeons and physicians until the fifteenth and sixteenth centuries (Siraisi 1990: 161).

23 Norri (1992: 296) reports a similar finding, although the difference between traditions is not as dramatic as in CEEM. Norri's published data does not show frequencies.

authors would use both within the same text, sometimes even within the same paragraph or sentence, as William Bullein did in *Dialogue Against the Feuer Pestilence* (1578):

(4) These are good **signes** general; but particular, what manifest **tokens** do signifie the Plague or Pestilence in a mannes owne proper bodie?[24]

Although it is possible for two equivalent terms to have been alternated merely to avoid repetition, little evidence would support such a notion – particularly in ME texts, where intratextual repetition was a regular feature. The medical signifiers *token* and *sign* were clearly interchangeable, but a closer look at collocations reveals something which may help us understand the guiding principles of at least some of the writers. Looking at *token* and *sign* in their immediate contexts, it appears that *token* was preferentially used for diagnostic signification while *sign* was preferred when prognosticating was involved.[25] While the differences are not stark enough to be in any way conclusive, they do suggest the possibility of an underlying reason. One explanation for this lexical phenomenon may lie in the contemporary collocations of *sign*. The lexical item appears frequently in medical (and other) texts as astrological sign – invariably referred to by *sign*, never by *token*. Astrology, which had a significant influence on medicine from the late Middle Ages to the end of the seventeenth century, was largely prognostic in nature.[26] As a testament

24 The quote includes the construction 'manifest token'. *Manifest* (= observable, apparent) appears in the CEEM corpus as a premodifier of signifier terms 13 times, the earliest attested occurrence in Thomas Phayer's *Book of Chyldren* (1546). The earliest OED quote with *manifest* as a premodifier is dated 1581 ('manyfest lyes'); the earliest quote in the medical context is dated 1951 ('manifest anxiety').

25 The findings reported here are tentative. A concordance analysis of the prognostic and diagnostic contexts of *token* and *sign* is in progress and exact figures are therefore not quoted.

26 Perhaps somewhat surprisingly, astrological medicine was most popular in the fifteenth and sixteenth centuries and not, as one might believe, in the Middle Ages. Early Modern medicine was greatly influenced by Arabic astrology, which introduced the new element of mathematical exactness to the study of the zodiac. The promise of greater accuracy in prognosis made this new learning appealing to many European doctors. Indeed, astrological studies were a part of medical studies at several universities around Europe during the

to this, 35% of the occurrences of *sign* (320 instances out of 915) had to be excluded from the study because they were used in an astrological sense. It would therefore make sense to presume that some writers may have attempted to avoid confusing their readers by reserving *sign* for this well-established context and using *token* for the more literal instances of signifying. By contrast, the corpus contains no texts where *accident* and *symptom* would appear to have been used as stylistic alternatives, suggesting that they were not interchangeable to the same degree.[27]

When we look at the contextual use of the signifier terms, it quickly becomes apparent that the interpretations of the Galenic theoretical difference between *sign* and *symptom* went through several stages. Up until the end of the sixteenth century, the conceptual separation between the two terms *token* and *sign* and the term *accident* appears to have been the most clear. The former were usually simple, indexical markers of ailments while the latter was only used when the referent was an ailment in itself. Consequently, *accident* could not refer to a medical indication which merely signaled something about an ailment, such as the color of a urine sample or the beat of the pulse. An intermediate period, ranging perhaps from around 1600 to around 1700, follows next, during which *accident* and *symptom* were understood by some physicians as a type of sign and by others as a co-occurring ailment, as before. As an example of the co-occurrent disease interpretation, de Vigo's *The Most Excellent Workes of Chirurgerye* (1543) describes *accident* in a manner very similar to the concept of complication:

(5) The syxth canon is, that ye muste not gyue a stronge laxatiue medicine, before flebothomye, for it myghte cause euyll **accidentes**.

Pomarius' *Enchiridion Medicum* (1609) appears to promote the opposite view, seeing accident as a type of sign:

---

late medieval and early Renaissance periods (Siraisi 1990: 68). Cf. Jones (1984: 67-75), Taavitsainen (1988).

27 The two terms do not appear within 100 words of each other in any of the corpus texts. Interestingly, the OED quotes for *symptom* list 3 examples in which they are used as an appositive pair.

(6) neither must we be ignorant, that sometimes a disease is not only essential, but **accidental** also, & a companio[n] of other diseases, so sometimes a feuer is an **accident** of many diseases, and of it selfe no disease: [...].

While awareness of the Galenic definitions of *sign* and *symptom* persisted well into the seventeenth century, it is evident that little by little the meanings of *accident* and *symptom* began to shift closer to *sign*. Banister defines *accident* explicitly as a diagnostic feature in *A needeful, new, and necessarie treatise of chyrurgerie* (1575):

(7) The differences of Ulcers are diversy [sic] taken, but especially of three causes, viz. from the causes prouoking them, frō the members, and from the **accidents** of dispositions ioyned with thē.

Of course, like all processes of semantic change, *accident*'s loss of the co-occurrent meaning did not happen overnight. *Accident* did not undergo a broadening of its signifier sense (as *symptom* did) and as the medical sense was gradually lost, *accident* began to disappear from medical use. It persisted as a medical term throughout the seventeenth century,[28] as illustrated by excerpt (8) from the *Philosophical Transactions* (vol. 18, no. 211, 1694):

(8) The Venom of a Viper in it self is not Mortal to a Robust and Sound Body; and tho very unhappy and mischievous **accidents** attend it; as a great Tumour, Tension and Weight of the Part, Humidity and Variety of Colours, Phrensies, Convulsions and Vomitings; yet in Eight or Ten days at most these **accidents** are over, [...]. [de Vaux: *A Discourse of the Viper* (1694)]

As the use of *accident* diminished, *symptom* took its place. Toward the end of the seventeenth century, *symptom* was on its way to becoming a well-established signifier both in medicine and, to a lesser degree, in the general lexicon. Brugis's *The marrovv of physick*e (1640) suggests that *symptom* is properly used in describing a physical ailment caused by a disease – see (9). While Brugis's language is notably similar to

28 Medical dictionaries serve as a good source on what terms were considered acceptable or up-to-date. *Accident*, despite its frequent use in medical writing, is not included at all as a headword in Garfield's *Physical Dictionarie* (1657) or in Quincy's *Lexicon physico-medicum* (1719); a headword does appear in Blankaart's *Physical Dictionary* (1684) but only in the form of a reference to *symptoma*.

classical Galenic descriptions of co-occurrence, he makes no reference to a secondary ailment and instead goes on to describe indexical conditions such as blindness and the formation of scabs as symptoms.

(9) There is (as I have shewed you before) three affects contrary to nature; the cause of a disease, the disease, and a **simptome**; the cause goeth before the disease, a **simptome** accompanies it, just as the shadow goeth with the body: in the generall signification it is called any thing whatsoever that chanceth to man contrary to nature, for whatever it be cause, or a disease, yet if they come in a disease and are over, and above nature, they are properly called **simptomes**.

This new, more general sense of *symptom* can be found in most dictionary definitions from the beginning of the eighteenth century. Instead of defining *symptom* as an exclusively medical term, many lexicographers treat it as a synonym of *sign*.[29] In medicine, this newly established synonymity threatened to render *symptom* redundant, a situation which later led to the modern practice of defining *sign* and *symptom* as objective and subjective signifiers respectively.[30]

Appositive word-pairs, a semantically interesting feature of early English, occur with some frequency with signifier terms in the corpus.[31] The most common type of doublet in vernacular medical writing involves the combination of a vernacular term with a roughly equivalent Latin or Greek lexical item, ostensibly for the purpose of either using one to clarify the meaning of the other or, in some cases, to more comprehensively cover a particular semantic field. Most signifier doublets (18 pairs out of a total of 22) in the corpus are of the *sign and token* variety, suggesting a clarifying function for the repetition since the two terms were semantic equivalents. Against expectations, however, these doublets occur almost exclusively in academic treatises (17 cases out of 22), a context which makes it

29 Examples include Kersey's *Dictionarium Anglo Britannicum* (1708) and Johnson's *Dictionary of the English Language* (1755).

30 The source of the first modern definitions for *sign* and *symptom* is unknown, although it is likely to be found in the late eighteenth or early nineteenth century when advances in laboratory science ushered in a new era of signs obtained by means of objective measurement. I am continuing research on this topic.

31 Cf. Koskenniemi (1968), McConchie (1997: 75-77), and Pahta / Nevanlinna (1997); see also Bugaj (this volume).

implausible that a doublet was used to define or clarify the meaning of *sign* with *token*. The more likely explanation is that the doublets were used as stylistic imitation of similar structures in Latin source texts and Latin writing in general. The few doublets of a different kind, usually the pair *sign and symptom*, appear to consist of clearly separate (if somewhat overlapping) semantic fields. *Token and accident* and *accident and symptom* do not appear in apposition at all. The *sign and symptom* pairs are used in describing the semiotic characteristics of diseases, both the primary features of the disease itself (signs) and the secondary co-occurring ailments (symptoms) which can also be used in identifying the primary disease but which are not signs in the stricter sense. Thus the formulaic expression *signs and symptoms* is not a case of repetition but rather quite necessary for a comprehensive description. In the centuries that follow, the coordinated noun pair *signs and symptoms* has become a formulaic, text organizing expression in medical writing, now indicating a comprehensive description of a different kind covering objective and subjective indications, respectively.

7. Conclusion

The results of this study, which combined a quantitative analysis of an electronic corpus with some qualitative findings from early dictionaries, demonstrated that both the usage and the semantic fields of the four terms have changed considerably over the period under investigation. The findings may be of interest to researchers of medical history, early lexicology, and historical pragmatics. In particular, the knowledge that the meanings of seemingly familiar terms like *sign* and *symptom* have in fact undergone quite radical changes could be of importance when earlier medical theories are compared to the current. Even more strikingly, the data showed that *accident*, a word which continues to be a regular part of the modern lexicon, had a frequent and significantly different meaning before the eighteenth century.

This study pointed out the need for further research into the syntactic structures of semiotic signification and the argumentation practices of medical semiotics. The usage patterns of the signifier terms need to be studied intertextually by tracing both their etymological sources and lines of transmission within the vernacular community. The fascinating history of *symptom* in English deserves to be looked at more closely, particularly in relation to the exact circumstances surrounding the redefinition of *sign* and *symptom* during the late eighteenth and early nineteenth centuries.

References

Primary sources (in order of citation)

(1) The Middle English Gilbertus Anglicus. Ms: Wellcome 537, ff. 48-310. In Getz, F.M. 1991. *Healing and Society in Medieval England. A Middle English Translation of the Pharmaceutical Writings of Gilbertus Anglicus*. Madison: Univ. of Wisconsin Press, 1-54.

(2) Lanfranc: Science of Cirurgie. Mss: Bodl. Ashmole 1396, ff 1r-272v; Bl Add 12,056, ff. 31r-86v. In Fleischhacker, R.V. 1894. *Lanfrank's* 'Science of Cirurgie', Part 1, *Text.* (Early English Text Society O.S 102). London: Kegan Paul, Trench, Trubner & Co., 7-58.

(3) Trevisa, John: On the Properties of Things. Ms: Bl Add. 27944. In Seymour, M.C. *et al.* 1975. *On the Properties of Things: John Trevisa's Translation of* Bartholomaeus Anglicus De Proprietatibus Rerum, *A Critical Text*, Vol. 1. Oxford: Oxford University Press, 163-196.

(4) Bullein, William 1578. *Dialogue Against the Feuer Pestilence.* STC (2nd ed.) / 4038. London: Iohn Kyngston.

(5) de Vigo, Joannes 1543. *The Most Excellent Workes of Chirurgerye.* London: Edwarde Whytchurche. Facsimile edn.

Amsterdam: Theatrum Orbis Terrarum and Da Capo Press, 1968. f. Ccxx-Ccxxv.
(6) Pomarius, Petrus 1609. *Enchridion Medicum: Containing an Epitome of the Whole Covrse of Physicke*. London: Henry Ballard for George Potter, 1609. STC (2nd ed.) / 24577, 1-55.
(7) Banister, John 1575. *A needeful, new, and necessarie treatise of chyrurgerie*. STC (2nd ed.) / 1360 London: Thomas Marshe, 1-17.
(8) de Vaux, Theodore 1694. A Discourse of the Viper, and some other Poysons, Wrote by Sr. Theodore de Mayerne, after Discoursing with Mr. Pontaeus. *Philosophical Transactions* 18/ 211, 162-166.
(9) Brugis, Thomas 1640. *The marrovv of physicke*. STC (2nd ed.) / 3931 London: Richard Hearne.

Secondary sources

Baer, Eugen 1981. Medical Semiotics: A New Paradigm. *Semiotica* 37, 169-191.
Eich, Wolfgang 1986. Eighteenth- and Nineteenth-century Medical Semiotics. *XXVII International Congress of the History of Medicine*, 503-505.
Foucault, Michel 1973. *The Birth of the Clinic: An Archeology of Medical Perception*. Sheridan-Smith, A.M. (transl.). London: Tavistock.
French, Roger 1994. Astrology in Medical Practice. In Garcia-Ballester, L. *et al.* (eds) *Practical Medicine from Salerno to the Black Death.* Cambridge: Cambridge University Press, 30-59.
French, Roger 2003. *Medicine before Science: The Business of Medicine from the Middle Ages to the Enlightenment.* Cambridge: Cambridge University Press.
Galenus, Claudius 2004. *Galen on Antecedent Causes*. Hankinson, R.J. (ed. and transl.). Cambridge: Cambridge University Press.
Garber, Marjorie B. 2001. *Academic Instincts*. Princeton: Princeton University Press.

Hess, Volker 1993. *Von der semiotischen zur diagnostischen Medizin. Die Entstehung der klinischen Methode zwischen 1750 und 1850*. Husum: Matthiesen.

Hess, Volker 1998. Medical Semiotics in the Eighteenth Century: A Theory of Practice? *Theoretical Medicine and Bioethics* 19/3, 203-213.

Hufeland, Christoph W. 1826. Über den Werth und die Bedeutung der Semiotik. *Journal der praktischen Heilkunde* 62/4, 3-9.

Jackson, Howard 2002. *Lexicography. An Introduction*. London: Routledge.

Jones, Claire 2004. Discourse Communities and Medical Texts. In Taavitsainen / Pahta (eds), 23-36.

Jones, Peter Murray 1984. *Medieval Medical Miniatures*. Austin: University of Texas Press.

Kersey, John 1969. *The New English Dictionary, 1702*. Menston: Scolar Press.

Koskenniemi, Inna 1968. *Repetitive Word Pairs in Old English and Early Middle English Prose. Expressions of the Type* Whole and Sound *and* Answered and Said, *and Other Parallel Constructions*. Turku: Annales Universitatis Turkuensis.

Landau, Sidney I. [2]2001. *Dictionaries: The Art and Craft of Lexicography*. Cambridge: Cambridge University Press.

Maloney, Gilles / Frohn, Winnie (eds) 1986-1989. *Concordance des œuvres hippocratiques*. Hildesheim: Olms-Weidmann.

McConchie, Roderick W. 1997. *Lexicography and Physicke*. Oxford: Clarendon.

Norri, Juhani 1992. *Names of Sicknesses in English, 1400-1550: An Exploration of the Lexical Field*. Helsinki: Suomalainen Tiedeakatemia.

Norri, Juhani 2004. Entrances and Exits in English Medical Vocabulary, 1400-1550. In Taavitsainen / Pahta (eds), 100-144.

Nutton, Vivian 1993. Galen at the Bedside: The Methods of a Medical Detective. In Bynum, F.W. / Porter, Roy (eds) *Medicine and the Five Senses*. Cambridge: Cambridge University Press, 7-16.

Pahta, Päivi / Nevanlinna, Saara 1997. Re-phrasing in Early English: Expository Apposition with an Explicit Marker from 1350 to 1710. In Rissanen, Matti / Kytö, Merja / Heikkonen, Kirsi (eds)

English in Transition: Corpus-based Studies in English Linguistics and Genre Styles. Berlin: Mouton, 121-183.

Pahta, Päivi 1998. *Medieval Embryology in the Vernacular. The Case of* De Spermate. Helsinki: Société Neophilologique.

Pahta, Päivi. 2004. Code-switching in Medieval Medical Writing. In Taavitsainen / Pahta (eds), 73-99.

Rausch, Irmengard 1999. *Semiotic Insights. The Data Do the Talking.* Toronto: Toronto University Press.

Risse, Günter B 1987. A Shift in Medical Epistemology: Clinical Diagnosis, 1770-1828. In Kawakita, Yosio (ed.) *History of Diagnostics.* Osaka: Div. of Medical History, The Taniguchi Foundation, 115-147.

Schäfer, Jürgen 1980. *Documentation in the O.E.D.* Oxford: Clarendon.

Schäfer, Jürger 1989. *Early Modern English Lexicography.* Oxford: Clarendon.

Sebeok, Thomas A 1976. *Contributions to the Doctrine of Signs.* Bloomington: Indiana University Press.

Sebeok, Thomas A 1999. *Signs: An Introduction to Semiotics.* Toronto: University of Toronto Press.

Sebeok, Thomas A. 2001. *Global Semiotics.* Bloomington: Indiana University Press.

Siraisi, Nancy 1990. *Medieval & Early Renaissance Medicine: An Introduction to Knowledge and Practice.* Chicago: University of Chicago Press.

Staiano, Kathryn Vance 1982. Medical Semiotics: Redefining an Ancient Art. *Semiotica* 38, 319-346.

Swales, John M. 1990. *Genre Analysis: English in Academic and Research Settings.* Cambridge: Cambridge University Press.

Taavitsainen, Irma 1988. *Middle English Lunaries: A Study of the Genre.* Helsinki: Société Neophilologique.

Taavitsainen, Irma 2004. Transferring Classical Discourse Conventions into the Vernacular. In Taavitsainen / Pahta (eds), 37-72.

Taavitsainen, Irma / Päivi Pahta 1998. Vernacularisation of Medical Writing in English: A Corpus-Based Study of Scholasticism. *Science and Medicine* 3, 157-185.

Taavitsainen, Irma / Pahta, Päivi 2004. Vernacularisation in Scientific and Medical Writing. In Taavitsainen / Pahta (eds), 1-22.

Taavitsainen, Irma / Pahta, Päivi (eds) 2004. *Medical and Scientific Writing in Late Medieval English.* Cambridge: Cambridge University Press.

Thagard, Paul 1999. *How Scientists Explain Disease.* Princeton: Princeton University Press.

Tyrkkö, Jukka 2006. Tokens, Signs, and Symptoms: Signifier Terms in Medical Texts from 1375 to 1725. In McConchie, Rod W. / Timofeeva, Olga *et al.* (eds) *Selected Proceedings of the 2005 Symposium on New Approaches in English Historical Lexis (HEL-LEX).* Somerville, MA: Cascadilla, 155-165.

Vila, Anna C. 2003. Reading the 'Sensible' Body: Medicine, Philosophy, and Semiotics in Eighteenth-century France. In Mitchell, Robert / Thurtle, Phillip (eds) *Data Made Flesh.* London: Routledge, 27-47.

Voigts, Linda Ehrsam 1982. Editing Middle English Medical Texts: Needs and Issues. In Levere, Trevor H. (ed.) *Editing Texts in the History of Science and Medicine.* New York: Garland, 39-68.

Voigts, Linda Ehrsam 1984. Medical Prose. In Edwards, A.S.G. (ed.) *Middle English Prose: A Critical to Major Authors and Genres.* New Brunswick, NJ: Rutgers University Press.

Appendix

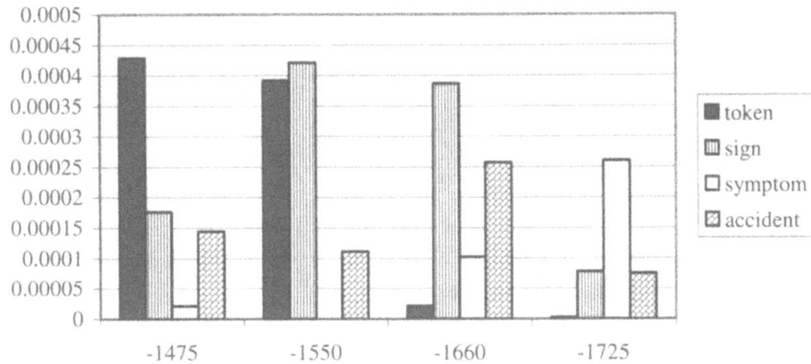

Figure 1. Frequencies of signifier terms.

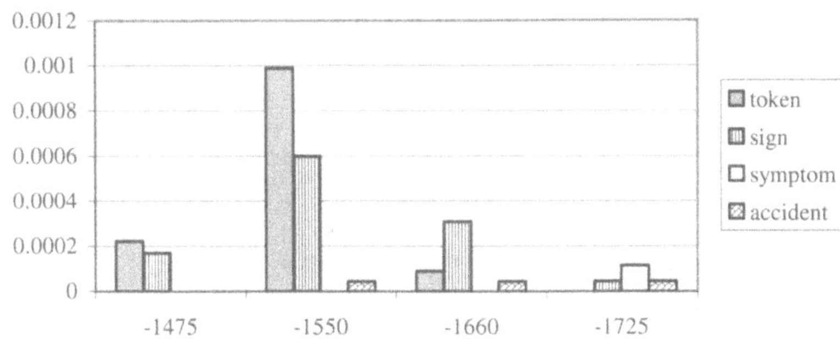

Figure 2. Signifier use in remedy texts.

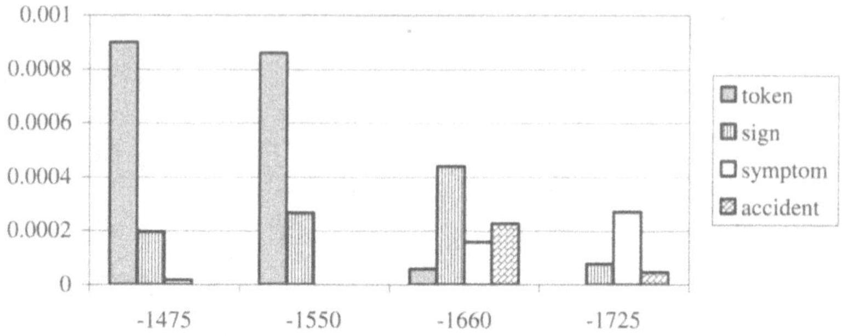

Figure 3. Signifier use in special treatises and *Philosophical Transactions* (1665-1712) (academic register).

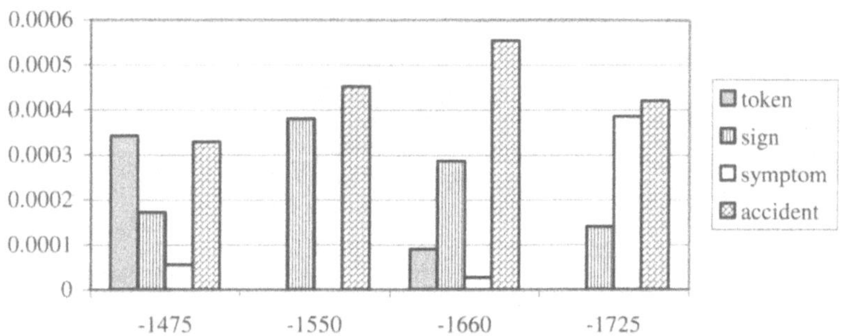

Figure 4. Signifier use in surgical texts (academic register).

ELENA SEOANE / CHRISTOPHER WILLIAMS

Changing the Rules: A Comparison of Recent Trends in English in Academic Scientific Discourse and Prescriptive Legal Discourse[1]

1. Introduction

The aim of this chapter is to compare the recent development of two types of specialized discourse in English, namely academic scientific discourse (ASD) and prescriptive legal discourse (PLD). Like most types of specialized discourse they are both characterized by a very high degree of formality and impersonalization (for an overview of the lexical, syntactic and textual features of specialized discourse, see Gotti 2003 and 2005). In many other ways, however, they are quite distinct.

What attracted us to make this comparison was the hypothesis that the recent evolution of these two types of specialized discourse might be following similar trends. As shown in Seoane (forthcoming), in contemporary ASD the use of the passive is declining markedly, particularly in American journals. Various factors are held responsible for this change, but the presence of what has been called a 'science-for-everyone' movement as possibly contributing to the shift towards a more accessible type of language seemed highly significant. As for PLD, Williams (forthcoming) describes the 'modal revolution' that has been taking place in certain parts of the English-speaking world in prescriptive legal texts, entailing the total abolition of that classic

1 Research for this study was funded by the Spanish Ministry of Education and Science (grant nos. BFF2001-2914 and HUM2004-00940) and the Galician Ministry of Education (grant nos. PGIDT01PXI20404PR and PGIDIT02-PXIC20402PN).

modal auxiliary of legal English *shall*[2] from certain laws, even from a constitutional Charter as in the case of South Africa, which was at least in part due to the influence of what is generally known as the Plain Language movement.

From these observations an obvious question emerges. If the passive – that iconic feature of scientific discourse – is fast being demonized in scientific style manuals with writers being encouraged to adopt active forms, and if *shall* – that equally stereotypical feature of prescriptive legal discourse – is likewise being ostracized in some quarters and replaced by other verbal constructions, could it be that the changes taking place in these genres are somehow linked? We decided to investigate the relationship between the two genres, trying to find out whether there could possibly be some kind of 'democratization' process occurring that underlies the changes taking place in each of the two genres or whether there are other factors involved. We were also interested in ascertaining if the changes were uniformly spread throughout the English-speaking world, and if there was any kind of resistance to change.

For this purpose, Sections 2 and 3 below concentrate on ASD and PLD respectively, and provide a brief outline of the major characteristics of each genre, followed by a look at some of the changes that have taken place in them over the years and what style manuals have to say on the subject. Section 4 offers an explanation for the differences observed in the recent evolution of these two types of discourse, and finally Section 5 provides the conclusions of this study.

2. Academic Scientific Discourse

2.1. The main features of ASD

As is well known, the Western scientific tradition from the nineteenth century is highly impersonal; the scientific author is expected to hide

2 Such was the predominance of *shall* in Early Modern English statutes that it constituted over 80% of all central modal auxiliaries. See Gotti (2001: 90-91).

behind a strictly objective, impersonal and neutral presentation of facts, and there is no room for originality in style. The linguistic indexes of this scientific style, which strives for objectivity and informational density, are passives, absence of personal pronouns, complex sentences, long noun phrases and nominalizations (Gopnik 1972; Gross 1985; Bazerman 1988; Swales 1990; Atkinson 1992 and 1999; Halliday / Martin 1993; Taavitsainen *et al.* 2002). These can be observed in (1), where passives appear in bold type and transitive actives in italics:

(1) When very high thermal efficiencies of, say, 95-98 per cent **are required**, a pure countercurrent formation **is invariably adopted**. Typical layouts **are shown** in Figures 9 and 10. The choice of headering **is governed** by several considerations, such as the operating pressures, the number of separate streams involved, and whether or not reversing duty **is included**. Figure 9 *shows* an arrangement suitable for two-stream steady duty, in which stream A is at comparatively low pressure. A more general solution **is shown** in Figure 10. This **is used** if streams A and B **are reversed** periodically, the geometry of these streams being symmetrical to maintain constant flow characteristics. Additional steady streams **may be included** as C, D, or E to suit requirements. This type of arrangement **is also used** when dealing with high pressure streams in all channels.
[A. G. Lemfesty, Low Temperature Heat Exchangers, in K. Mendelsohn (ed.), 1961, *Progress in Cryogenics*, 3. London: Heywood,: 28-33 (ll. 68-81)]

Scientific English in the nineteenth century, therefore, shifts away from an author-centred rhetoric to an abstract form of language, mainly with the purpose of conveying a greater sense of objectivity and scientific validity. It reaches a high level of abstraction and unintelligibility, which leads scholars to label the process of adapting a scientific article for a wider audience as a "scientific translation for the laity," or "metamorphosis of a technical text for the general audience" (Halliday / Martin 1993: 180).

2.2. Signs of change in ASD

A recent and ongoing change observed in scientific British and American English is the dramatic decrease in the frequency of the use of the passive voice. In Seoane (forthcoming) four matching corpora from the twentieth century were selected, which are designed with the explicit aim of comparing them with one another. They are the LOB and Brown, containing standard British and American English published in 1961, and the FLOB and Frown, which are matching samples from the early 1990s. This paper is based on a larger corpus, both in size, with 100,000 new words, and also chronologically, as can be seen in Table 1.

	Period I: early twentieth c. (1905 / 1925)	Period II: mid twentieth c. (1960 / 1975)	Period III: late twentieth c. (1985 / 1990)
British E.	30,000	40,000	40,000
American E.	30,000	40,000	40,000

Table 1. Number of words per subperiod.

These texts are all research articles published in scientific journals; most of them have been extracted from ARCHER (Biber / Finegan / Atkinson 1994) and others from the Internet. In all three periods two-thirds of the articles deal with the Natural Sciences, the other third with Medicine.

Table 2 displays the number of actives and passives found (also provided in graph form in Figure 1). The passives analysed in this paper are Quirk *et al.*'s (1985: § 3.75) "central" *be*-passives occurring in finite verb groups. As for actives, we have only considered those standing in a relation of variation with the passive, that is, active transitive clauses with an overt object eligible to become a passive subject, actives for which a passive counterpart would be available. In Table 2 and the corresponding Figure 1, the following developments become apparent. Firstly, there is a great deal of continuity in the results found for the beginning of the twentieth century, with no difference in the proportion of passives in British English and only a two point difference in American English. Secondly, there is a pronounced decrease in the use of passives in the second half of the century. In British English we observe a decrease from 66.4% to 58.7% of all transitive clauses. This decrease is more pronounced in the case of American English, from 64.2% to 46.4%, which means that the predominance of passives over actives is now reversed.

Periods		*I*		*II*		*III*	
Verb forms		*Actives*	*Passives*	*Actives*	*Passives*	*Actives*	*Passives*
Br	No	405	801	548	1,084	554	788
E	%	33.5%	66.4%	33.5%	66.4%	41.2%	58.7%
Am	No	517	930	614	1012	848	736
E	%	35.7%	64.2%	37.7%	62.2%	53.5%	46.4%

Table 2. Number of transitive active and passive verb phrases.

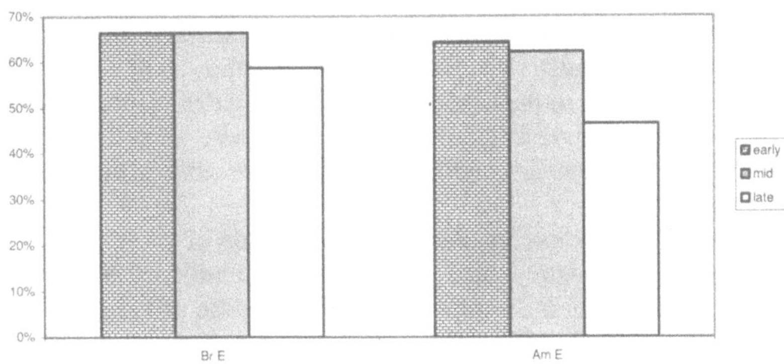

Figure 1. Actives and passives in scientific British and American English: early twentieth c., mid-twentieth c., and late twentieth c.

This reversal can be interpreted as a syntactic shift of markedness, since passives have become the new minority form in scientific American English and are therefore the pragmatically and stylistically marked form (Rodman 1994). The following may be considered a typical example, where actives and passives alternate naturally as in other types of genres:

(2) To compare measurements by investigators in different laboratories and apply their results to atmospheric or astrophysical measurements, *we must also understand* the corrections applied for the isotopic composition of the measured gas samples. Unfortunately, in many papers, the isotopic composition **is not specified**, and the interpretation of 'natural abundance' **is left** to the reader. Johns [8] *discussed* in detail the difference between the 'natural abundance' of 13C in atmospheric CO2 and that in commercially supplied CO2. In our conversion of reported intensities to a consistent set of units, we *have attempted* to rescale each intensity value to represent the band intensity for a 100% pure sample of a single isotope. Where no specific

> isotopic abundance information **has been given** in the original reference, we *have used* the 1986 HITRAN abundances.
>
> [K. Narahari Rao / A. Weber 1992. *Spectroscopy of the Earth's Atmosphere and Interstellar Medium.* Boston: Academic Press, 154-158 (ll. 86-100)]

2.3. ASD style manuals

Manuals of scientific style have traditionally recommended shunning passives on the grounds that they are weak, ambiguous, pedantic, longwinded, or deadly. The famous maxim "thou shalt not use the passive" dates back to the first quarter of the twentieth century (Strunk 1918), though statistical data reveal that passives were nevertheless predominant in scientific English throughout the century, until the last two decades.

However, a close examination of manuals of style shows that there has been a change which can be dated around the 1980s. Before this time manuals of style either do not mention the use of passives at all (e.g. Baker 1961; Hill 1971) or recommend using actives instead on the grounds that passives are stylistically longwinded and tedious (Earle 1911; Westaway 1926; Trelease 1951; Zall 1964; Sherman 1966; Mitchell 1967; Weisman 1968; Turner 1974). Only two style manuals have been found to offer a functional explanation for the use of the passive in scientific English before the 1980s, namely Cooper (1964) and *Scientific Style and Format: The CBE Manual for Authors, Editors, and Publishers* (1972); in these works the authors clarify when and why passives should be used, for reasons of emphasis, clause linkage and impersonalization. After the 1980s, however, although many manuals still omit mention of the passive (Woodwark 1992; Blum / Knudson 1997) or refer to the passive only cursorily (Farr 1985; Katz 1985; Bolsky 1988; Seely 1998; Hancock 2003), manuals which contain a detailed analysis of the passive abound, pointing to fully justified functional factors for their use (Alvarez 1980; Andrews / Blickle 1982; Master 1986; Wilkinson 1991; Lindsay 1995; Rubens 2001; Alley 2003). In these manuals more emphasis is placed on the role of the text author, with clarity and intelligibility as the desired aims, while pompous language and elaborate grammar are considered detrimental to the stylistic quality. This functional justification for the use of passives is new in manuals of style and is a clear indicator that the change is on-going and was spreading within the scientific community at the end of the twentieth century.

3. Prescriptive Legal Discourse

3.1. The main features of PLD

As Gibbons (1999) has observed, the law "involves the imposition of societal norms, associated with the deployment of power." PLD thus includes all those legal written documents which regulate social situations and behaviour. It not only includes laws and regulations passed by authorities which may be municipal, federal, national or international, and so on, but also contracts and wills. Here, for reasons of space, we shall be focusing exclusively on laws, directives and regulations.

The specificities of PLD in English include: the frequent repetition of particular words, phrases and syntactic structures; long, complex sentences with a high degree of subordination and co-ordination; syntactic discontinuities; the frequent use of passive constructions; and an impersonal style (Gotti 1991; Bhatia 1993; Tiersma 1999; Williams 2005). Instances of all these features can be found in the following 120-word sentence:

(3) Powers of vice-chair 11. Where (a) a member of a Board is appointed to be vice-chair either by the Assembly or under regulation 10, and (b) the chair of the Board has died or has ceased to hold office, or is unable to perform the duties of chair owing to illness, absence from England and Wales or any other cause, the vice-chair shall act as chair until a new chair is appointed or the existing chair resumes the duties of chair, as the case may be; and references to the chair in Schedule 3 shall, so long as there is no chair able to perform the duties of chair, be taken to include references to the vice-chair.

[Section II of the *Local Health Boards (Constitution, Membership and Procedures) (Wales) Regulations* 2003)]

3.2. Signs of change in PLD

The results in this section are based on six small corpora, specially compiled for this study, consisting of prescriptive legal texts in English of approximately 20,000 words each (the smallest is made up of 20,161 words, the largest of 22,487 words) coming from three different sources, namely the US, New South Wales, and the European Union, and grouped into two periods of time – 1970-76 and 2003-05 – thus making it possible to discern changes over a thirty-year period. Every attempt was made to ensure that the data from each source were roughly comparable. Hence, for example, in the case of European Union texts, only EC directives were taken into consideration for both periods of time. As with the results on passives in ASD outlined above in Section 2.2, the only active forms taken into consideration were transitive clauses with an overt object eligible to become a passive subject.

Unlike the situation in ASD, there would not appear to be a general drop in the frequency of passives in PLD in recent decades. Indeed, in some cases there has been a tendency towards an increase in passive use over the last 30 years. For example in the legislative texts from the US there has been a rise from 41.1% in the 1970s to 49.7% in recent texts. In EC directives drafted in English passives have risen marginally from 56.8% in texts from 1972-1975 to 57.7% in texts from 2003-2005. On the other hand, if we take comparable corpora, for example, from New South Wales, the picture is rather different, with a drop in passives from 63.7% of all transitive verbal constructions in the early 1970s to 45.8% in recent texts. However, it should also be pointed out that the frequency of passives in the NSW texts of the 1970s was exceptionally high, almost comparable to that of old-style scientific texts.

As for the use of *shall*, there has been nothing less than a revolution in legislative texts in English from the southern hemisphere as a result of the decision to eliminate it entirely from texts drafted in Australia, New Zealand and South Africa. One case that is sometimes cited (e.g. Williams forthcoming) – because it is possible to make a direct comparison with the situation just three years beforehand – is the difference between the *South Africa Interim Constitution* of 1994 which contains 1288 cases of *shall* and 0 cases of *must*, whereas the definitive text of the *Constitution* drafted less than three years later contains 0 cases of *shall* and 414 cases of *must*. Of course, the jettisoning of *shall* is only one of a number of reforms in the drafting of legislative texts (for detailed information and suggestions about modern legal drafting techniques in line with Plain Language principles seen from the perspective of a government body, see Australian Government Office of Parliamentary Counsel 2005).

However, we should be wary of generalizing these admittedly striking results. If we shift our attention towards legislative texts emanating from the major international organizations such as the United Nations, the European Union or the International Labour Organization, the picture would seem to be one of continuity and stability through the years, at least as regards the frequency of *shall*.[3] To give just one example, according to data provided by Véronis (2005) the English version of the European Constitution in its entirety contains 3190 instances of *shall* and recurs approximately once every 40 words. It is the sixth most commonly used word in the text after *the*, *of*, *and*, *to* and *in*, followed by *a*, *article*, *European* and *be*. Returning to our data on EC directives, we actually find a slight increase in cases of *shall* over the years, rising from 25.4% of all finite verbal constructions in the texts from 1972-1975 to 28.4% in texts from 2003-2005. In the United States, however, we see a drop from 38.3% to 25.2% (see Figure 2). But again, as with the 1970s data on passives for New South Wales, we need to bear in mind that the 1970s figure is exceptionally high.

In general terms, however, if we look at PLD in English at a world level there would seem to be a fairly clear north-south divide, with the southern hemisphere displaying a far greater willingness to innovate compared with the more conservative northern hemisphere. Of course, *shall*-free texts could be seen as being the vanguard of a trend that will eventually sweep through the entire English-speaking world. Alternatively, this dual system may carry on as it is for decades – even centuries – to come. Both scenarios are plausible.

---

3 Authentic legislative texts in English can be found online, respectively, at <www.un.org>, <europa.eu.int/index_en.htm>, and <www.ilo.org>.

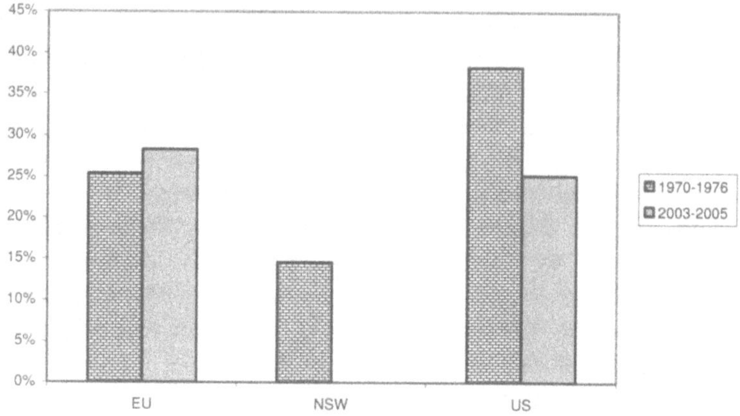

Figure 2. Comparative distribution of *shall* in legal texts from the European Union, New South Wales, and the United States: 1970-1976 and 2003-2005.

3.3. PLD style manuals

As with ASD, there are a number of manuals offering suggestions and guidelines for drafters of legislative texts. But whereas science style manuals would seem to offer substantially similar advice throughout the English-speaking world, the situation of legal style manuals is much more variegated. This may be partly due to the fact that the criteria for writing an academic scientific article are relatively uniform, while there may be considerable differences in the criteria for drafting, say, a piece of local legislation with respect to a UN Security Council resolution. This in turn depends to some extent on the expected readership of each type of discourse: accredited scientific journals are meant to diffuse within the international scientific community whereas the intended readership of legal texts may sometimes be more local. Each national legal culture follows its own conventions which tend to be relatively impervious to change or to imitation of other models. This is also true of the drafting of international legal texts: the UN, the EU and the ILO each has its own well-established drafting conventions. And once those conventions have been decided on and been found to work, drafters and all those involved in the law-making process are generally loath to introduce sweeping stylistic changes.

One particularly emblematic case concerns the 'tug-of-war' within the EU between those who wish to blow away "that vague grey pall that descends on EU documents, obscuring meanings and messages, causing delays and irritation" (Fight the Fog Campaign 2005) and those EU officials who fear the consequences and possible confusion that might ensue if there is a radical change in the way legally binding texts are drafted. Using the approach that is familiar to Plain Language practitioners of 'before' and 'after' – i.e. first providing the original text then reformulating it along more user-friendly lines – Martin Cutts, in his booklet *Clarifying Eurolaw* (2001), has shown how the EU directive 88/378/EEC on the safety and free trade of toys could be rewritten, with its layout partly revised, and a 'citizen's summary' added. As is outlined in Cutts / Wagner (2002: 4-5), a thousand copies of the booklet were distributed to European Commission administrators who produced the first drafts of EU legislation, but the "European Commission's legal service responded unfavourably saying that '[the] use of archaic language and legalese is not a serious problem in Community legislation'." Moreover, it goes on: "The legal service saw as a 'false premiss' the booklet's statement that 'legislators should regard the primary audience of every directive as interested private citizens, not lawyers'" (Cutts / Wagner 2002: 5). So far it would appear that, within the EU, the forces resisting major stylistic innovations in legal drafting still predominate, but the calls for change are growing too, also from those working within the EU, as is testified to by the activities of the Fight the Fog Campaign.

Taking a worldwide perspective, if we look at the burgeoning number of contemporary legal drafting manuals in English, we can divide them roughly into four basic categories, namely, those which are written by:

i) 'independent' authors who identify themselves as being part of or close to the Plain Language movement (e.g. Wydick 1994; Dick 1995; Maher 1996; Close 1998; Mowatt 1999; Cutts 2000 and 2001; Butt / Castle 2001; Garner 2001; Asprey 2003; Aitken / Butt 2004);

ii) 'independent' authors who do not necessarily identify themselves explicitly with the principles of the Plain Language movement, even if their works may contain elements and features which coincide with Plain Language principles (e.g. Charrow / Erhardt 1986; Kessler 2002; Kessler / Hunter 2003; Adams 2001 and 2004);

iii) official bodies which explicitly embrace the principles of the Plain Language movement (e.g. Law Reform Commission of Victoria 1987; Office of Investor Education and Assistance U.S. Securities and Exchange Commission 1998; New South Wales Government Parliamentary Counsel 2000; Australian Government Office of Parliamentary Counsel 2005; Maine Legislative Drafting Manual 2005; New Zealand Parliamentary Counsel Office 2005);

iv) official bodies which do not specifically adhere to the principles of the Plain Language movement, though, as with ii), their works may contain elements and features which coincide with Plain Language principles (e.g. International Labour Organization 2005; North Dakota Legislative Branch 2005; Texas Legislative Council Drafting Manual 2005).

4. Why the differences?

In this section we try to account for the differences in the rates and type of change between ASD and PLD by addressing questions such as why it is that the editors of and contributors to American – and to a considerable extent also British – scientific journals have willingly embraced the idea of stylistic change, while the language of most American, British and EU legislative texts has proved to be much more resistant to change, especially if compared with what has been happening at the southern end of the English-speaking world such as Australia, New Zealand and South Africa. To answer this question we will first put forward two hypotheses as to why ASD in English seems to be changing on a worldwide level.

The first hypotheses considered in Seoane (forthcoming) for the decrease in use of passives are purely linguistic, namely the alleged colloquialization of written English or the fact that scientific discourse may be becoming more personal or more subjective. The evidence provided there would seem to suggest that this is not the case and that other factors, of a socio-cultural nature, determine the change.

One socio-cultural factor playing a decisive role in the decrease of passives is the need for scientists to succeed in an increasingly competitive market since, according to some estimates, the number of scientific publications doubles every ten to fifteen years (Gläser 1995: 58). Scientists must secure a publication source for their findings, and must therefore convince an editorial board that their work will have a diffuse readership. Diffusion largely depends on accessibility and clarity, and, as a consequence, traditional attitudes like the excessive use of conventionalized passives are being called into question because they are inefficient. This is why style manuals, especially after the 1980s, concur that the writer/scientist must not shun all passives but only the passives which have no linguistic function. In manuals of style such passives are referred to as "unnecessary passives" (Zetter / Crouch 1961) and as the "thoughtless use of passive" (Mills / Walter 1978). Indeed, a close analysis of the passives occurring in most early twentieth century scientific texts will show that, in many cases, they have no evident linguistic function. Indeed, passives are claimed to have lost the communicative function they originally had in scientific writing, which was that of conveying a sense of objectivity. They became conventionalized, extended and generalized, and they lost their primary function. Like any linguistic feature which does not have a communicative function in informative writing, conventionalized passives are destined to be deleted (Halliday / Martin 1993: 10). Therefore, one factor determining the decay of passives is the fact that they have become mere conventional ornaments and are functionally empty; they are a waste of time and energy for the reader, and in the modern world the reader has less and less time and energy to waste. As the President of the Royal Society put it: "Anyone who writes in this [passive] style today simply is not likely to be at the cutting edge" (May 2003).

Secondly, the decay in the use of passives also seems to derive from a reaction against the detached and alienating style characteristic of nineteenth-century and early-twentieth-century scientific discourse. This reaction stems from a new way of understanding science which aims to include everyone, not only the experts, without projecting this barrier between the message and the reader that stiff language constituted in the past. In Halliday / Martin's words (1993: 53), "Scientific language [...] creates a massive disjunction between everyday commonsense knowledge and the systematized knowledge of the disciplines." The new scientific approach no longer seeks to be marked off as the discourse of an intellectual elite but, on the contrary, it wants to be tolerant with a commonsense approach to science which reflects scientists' current perceptions. Several authors acknowledge a marked tendency towards more democratic forms of discourse, in tune with what has been called the 'democratization of discourse' by which all the linguistic features expressing inequalities of power "become less marked as society itself becomes, in appearance at least, more democratic" (Smith 2003: 253).

As for PLD, let us now hypothesize why legal drafters in the southern hemisphere have generally been more willing than their northern counterparts to accept or promote radical stylistic changes. The former evidently feel less the weight of tradition, and hence are more prepared to experiment, whereas the legal culture which still predominates in the US and in particular the UK would seem to be far more respectful of tradition and of continuity. One eloquent example is the so-called enactment clause which we find at the start of each piece of legislation. Here is the standard formula to be found in all laws, local or national, passed in the UK:

(4) Be it therefore enacted, by the Queen's Most Excellent Majesty, by and with the advice and consent of the Lords Spiritual and Temporal, and Commons, in this present Parliament assembled, and by the authority of the same, as follows.

In the US, South Africa and New Zealand the formula has been substantially simplified, but it still retains a decidedly archaic quality through the use of the passive imperative construction (the jussive subjunctive) 'Be it [...] enacted by [...]' etc. In Canada the active form is used, whereas the formula used by the Parliament of South Australia could hardly be more straightforward:

(5) The Parliament of South Australia enacts as follows.

The propensity to accept and introduce changes in the language of the law is linked to the degree of permeability of the legal hierarchy in the countries and international organizations examined. To give another eloquent example, the decision not to include *shall* in the South African Constitution of 1997, i.e. at a time of epoch-making changes in South Africa's history, was the direct result of appointing onto the Constituent Assembly a Canadian lawyer, Philip Knight, whose views as an exponent of the Plain Language movement were well-known. Without the presence of this one individual it is likely that the Constitution would have been written in the traditional style. But a break with the past, with its memories of apartheid and discrimination, was evidently considered necessary not only politically but also linguistically. On the contrary, one wonders whether it even occurred to anyone involved in the setting up of the newly devolved assemblies of Scotland, Wales or Northern Ireland in the late 1990s to find a new type of language for drafting legislation, just as it would be hard to imagine the appointment of Plain Language representatives from various countries onto the Constituent Assembly of the European Constitution. However, perhaps after the negative results of the referendums in France and Holland in 2005 there may be a few misgivings at not having produced a document that most citizens would find more user-friendly. Continuity within a traditional framework would seem to be the underlying criterion uniting the logic of the powers-that-be in the US, the UK and the major international organizations. Some of the lexis may change and be updated (though by no means all of it), but the underlying concepts continue to be encased in grammatical structures that have evolved but slowly if at all over the decades, using formulae that have withstood the test of time.

The general tendency in PLD not to reduce the frequency of the passive, in marked contrast with what has been happening in ASD, would seem to be highly significant. If one looks through legal style manuals, particularly those written by Plain Language exponents, the advice to drafters to reduce the number of passive constructions is reiterated almost like a mantra, frequently accompanied by that famous citation by George Orwell (1946) "Never use the passive where you can use the active." But of course Orwell was referring to other types of written discourse such as journalism and prose, not to prescriptive legal texts. In fact, many legal drafting manuals fail to give concrete examples from *legal* texts in order to justify the call to shy away from the passive form. Not surprisingly, then, drafters of laws would seem to have ignored this advice. In other words, the suggestion to avoid passives in legal drafting manuals is essentially rhetorical. In practice the passive in PLD has been, and continues to be, an extremely effective device for conveying certain types of obligations and rights, generally in an impersonalized way which befits PLD. However, as the data from New South Wales texts have shown, it is possible to reduce an excessive use of the passive form.

Although the passive construction has been declining in use in British English in general over the last few decades, most of today's legal drafters are probably unconcerned if the text sounds a little antiquated or stilted: the overriding priority is that it does its job efficiently and can withstand the scrutiny of lawyers bent on finding loopholes and ambiguities.

As mentioned earlier, a scientific journal cannot afford to look outdated in terms of style: it must be seen as representing the 'cutting edge' of academic knowledge and expertise. Not so most legislative texts in English – at least, not from the northern hemisphere which, even today, may still contain a sprinkling of *hereinafters* and other relics of a bygone era without a twinge of self-consciousness. Where revolutionary changes in style have taken place in PLD, it has been largely the result of pressure from groups who have been given credibility and trust by the legal discourse community.

5. Conclusion

The decrease in *be*-passives in scientific English exemplifies how genres, even those which are strongly conventionalized like the scientific genre, may evolve along with changing historical conditions. Scientific English exhibits changes which respond to external socio-cultural influences, namely the democratization of society and the pressure to convey scientific knowledge as efficiently and clearly as possible. On the other hand, the reluctance of many legal drafters to incorporate stylistic changes, despite the pressure of groups such as the Plain Language movement, is a reflection of the essentially conservative nature of legal systems the world over. In the end, the difference in the respective rates of change is also a reflection of the difference in power status: academics may more readily adapt to changes, consciously updating the language they use to convey their ideas; on the contrary lawmakers still tend to impose changes using the language of authoritativeness and tradition. And, as we all know, it is the latter who wield true power!

References

Primary Sources

Adams, Kenneth A. 2001. *Legal Usage in Drafting Corporate Agreements*. Westport CT: Quorum Books.

Adams, Kenneth A. 2004. *A Manual of Style for Contract Drafting*. Chicago: ABA Press.

Aitken, James K. / Butt, Peter 2004. *Piesse: The Elements of Drafting*. Lawbook Co.

Alley, Michael 2003. *The Craft of Scientific Presentations*. Chicago: University of Chicago Press.

Alvarez, Joseph A. 1980. *The Elements of Technical Writing*. New York: Harcourt Brace: Jovanovich.

Andrews, Deborah C. / Blickle, Margaret D. [2]1982. *Technical Writing: Principles and Forms*. New York: Macmillan.

Asprey, Michèle 2003 (1991). *Plain Language for Lawyers*. Annandale, NSW: The Federation Press.

Australian Government Office of Parliamentary Counsel 2005. At http://www.opc.gov.au/index.htm.

Blum, Deborah / Knudson, Mary 1997. *A Field Guide to Science Writers*. Oxford: Oxford University Press.

Bolsky, Morris I. 1988. *Better Scientific and Technical Writing*. Englewood Cliffs, NJ: Prentice Hall.

Butt, Peter / Castle, Richard 2001. *Modern Legal Drafting. A Guide to Using Clearer Language*. Cambridge: Cambridge Univ Press.

Charrow, Vedra R. / Erhardt, Myra K. 1986. *Clear and Effective Legal Writing*. Boston: Little / Brown.

Close, Arthur L. 1998. Gender-free Legal Writing. British Columbia Law Institute. At http://www.bcli.org/pages/projects/.

Cooper, Bruce M. 1964. *Writing Technical Reports*. London: Penguin.

Cutts, Martin / Wagner, Emma 2002. *Clarifying EC Regulations*. Stockport: The Plain Language Commission Peak Press.

Cutts, Martin 2000 (1994). *Lucid Law*. Stockport: The Plain Language Commission Peak Press.

Cutts, Martin 2001. *Clarifying Eurolaw*. Stockport: The Plain Language Commission Peak Press. At <europa.eu.int/comm/translation/en/ftfog/eurolaw/eurolaw.htm>.

Dick, Robert C. 1995. *Legal Drafting in Plain Language*. Scarborough, Ontario: Carswell Press.

Earle, Samuel Chandler 1911. *The Theory and Practice of Technical Writing*. New York: Macmillan.

Farr, Alfred D. 1985. *Science Writing for Beginners*. Oxford: Blackwell.

Fight the Fog Campaign 2005. At <europa.eu.int/comm/translation/en/ftfog/>.

Garner, Bryan 2001. *Legal Writing in Plain English*. Chicago: University of Chicago Press.

Hancock, Elise 2003. *Ideas into Words: Mastering the Craft of Science Writing*. Baltimore: John Hopkins University Press.

International Labour Organization 2005. *Labour Legislation Guidelines*. At <www.ilo.org.public/english/dialogue/ifpdial/llg /index.htm>.

Katz, Michael J. / Massey, Harrie 1974. *General Notes on the Preparation of Scientific papers*. London: The Royal Society.

Kessler, James / Hunter, Fiona 2003. *Drafting Trusts and Will Trusts in Canada*. Markham: LexisNexis Butterworths.

Kessler, James 2002. *Drafting Trusts and Will Trusts: A Modern Approach*. London: Sweet / Maxwell.

Law Reform Commission of Australia 1987. *Plain English and the Law*. Melbourne: Victorian Government Printer.

Lindsay, David 1995. *A Guide to Scientific Writing*. London: Longman.

Maher, Chrissie 1996. *Language on Trial: The Plain English Guide to Legal Writing*. London: Robson.

Maine Legislative Drafting Manual 2005. At <janus.state.me.us/legis/ ros/manual/Webdman12.htm#P2015_149587>.

Master, Peter A. 1986. *Science, Medicine and Technology: English Grammar and Technical Writing*. Englewood: Prentice-Hall.

Mills, Gordon H. / Walter, John A. [4]1978. *Technical Writing*. New York: Holt.

Mitchell, John 1967. *A First Course in Technical Writing*. London: Chapman and Hall.

Mowatt, Christine 1999. *A Plain Language Handbook for Legal Writers*. Scarborough, Ontario: Carswell.

New South Wales Government Parliamentary Counsel [8]2000. *Manual for the Preparation of Legislation*. At <www.pco.nsw.gov.au/ pdf/mpl.pdf>.

New Zealand Parliamentary Counsel Office 2005. At <www.pco. parliament.govt.nz/index.shtml>.

North Dakota Legislative Branch 2005. *Legislative Drafting Manual*. At <www.state.nd.us/lr/information/bills/draft-manual.html>.

Office of Investor Education and Assistance U.S. Securities and Exchange Commission 1998. *A Plain English Handbook: How to Create Clear SEC Disclosure Documents*. At <www.sec.gov/ pdf/handbook.pdf>.

Rubens, Philip [2]2001. *Science and Technical Writing. A Manual of Style*. New York: Routledge.

Scientific Style and Format: The CBE Manual for Authors, Editors, and Publishers. [3]1972. Cambridge: Cambridge Univ. Press.

Seely, John 1998. *The Oxford Guide to Writing and Speaking: The Key to Effective Communication*. Oxford: Oxford Univ. Press.

Sherman, Theodore Allison [2]1966. *Modern Technical Writing*. Englewood: Prentice Hall.

Strunk, William 1918. *Elements of Style*. Ithaca, NY. Privately printed.

Texas Legislative Council Drafting Manual 2005. At <www.tlc.state.tx.us/legal/dm/sec729.htm>.

Trelease, Sam F. 1951. *The Scientific Paper, How to Prepare It, How to Write It: A Handbook for Students and Research Workers in All Branches of Science*. Baltimore, MD: Williams / Wilkins.

Turner, Barry T. 1974. *Effective Technical Writing and Speaking*. London: Business Books.

Véronis, Jean 2005. Technologies du langage. Text: The Words of the European Constitution. At <aixtal.blogspot.com/2005/04/ text-words-of-european-constitution.html>.

Weisman, Herman M. 1968. *Basic Technical Writing*. Columbus, OH: Merrill.

Westaway, Frederic W. 1926. *The Writing of Clear English. A Book for Students of Science and Technology*. London: Blackie.

Wilkinson, Antoinette Miele 1991. *The Scientist's Handbook for Writing Papers and Dissertations*. Englewood: Prentice Hall.

Woodwark, John 1992. *How to Run a Paper Mill: Writing Technical Papers and Getting Them Published*. Winchester: Information Geometers.

Wydick, Richard C. 1994. *Plain English for Lawyers*. Durham, NC: Carolina Academic Press.

Zall, Paul Maxwell 1964. *Elements of Technical Report Writing*. New York: Harper.

Zetter, Robert L. / Crouch, William G. 1961. *Successful Communication in Science and Industry Writing*. New York: McGraw-Hill.

Secondary Sources

Atkinson, Dwight 1992. The Evolution of Medical Research Writing from 1735 to 1985: The Case of the Edinburgh Medical Journal. *Applied Linguistics* 13/4: 337-374.

Atkinson, Dwight 1999. *Scientific Discourse in Sociohistorical Context. The Philosophical Transactions of the Royal Society of London, 1675-1975*. London: Erlbaum.

Bazerman, Charles 1988. *Shaping Written Knowledge. The Genre and Activity of the Experimental Article in Science*. Madison: Univ. of Wisconsin Press.

Bhatia, Vijay K. 1993. *Analyzing Genre: Language Use in Professional Settings*. London: Longman.

Biber, Douglas / Finegan, Edward / Atkinson, Dwight 1994. ARCHER and its Challenges: Compiling and Exploring a Representative Corpus of Historical English Registers. In Fries, Udo / Tottie, Gunnel / Schneider, Peter (eds), *Creating and Using English Language Corpora.* Amsterdam: Rodopi, 1-14.

Gibbons, John 1999. Language and the Law. *Annual Review of Applied Linguistics* 19: 156-173.

Gläser, Robert 1995. *Linguistic Features and Genre Profiles of Scientific English.* Frankfurt am Main: Peter Lang.

Gopnik, Myrna 1972. *Linguistic Structures in Scientific Texts*. The Hague: Mouton.

Gotti, Maurizio 1991. *I linguaggi specialistici.* Florence: La Nuova Italia.

Gotti, Maurizio 2001. Semantic and Pragmatic Values of *shall* and *will.* In Gotti, Maurizio / Dossena, Marina (eds), *Modality in Specialized Texts*. Bern: Peter Lang, 89-111.

Gotti, Maurizio 2003. *Specialized Discourse. Linguistic Features and Changing Conventions*. Bern: Peter Lang.

Gotti, Maurizio 2005. *Investigating Specialized Discourse*. Bern: Peter Lang.

Gross, Alan G. 1985. The Form of the Experimental Paper: A Realization of the Myth of Induction. *Journal of Technical Writing and Communication* 15: 15-26.

Halliday, Michael A.K. / Martin, James R. 1993. *Writing Science: Literacy and Discursive Power*. Pittsburgh: Univ. of Pittsburgh Press.

Maley, Yon 2001 (1994). The Language of the Law. In Gibbons, John (ed.), *Language and the Law*. London: Longman, 11-50.

May, Robert 2003. *'I did' or 'It was done'. Some Thoughts on Scientific Writing Style. Osmosis* 42.

Orwell, George 1946 (1954). Politics and the English Language. In *A Collection of Essays*. New York: Doubleday Anchor Books, 167-177.

Rodman, Lilita 1994. The Active Voice in Scientific Articles. Frequency and Discourse Functions. *Journal of Technical Writing and Communication* 24/3: 309-331.

Seoane, Elena forthcoming. Changing Styles: On the Recent Evolution of Scientific British and American English. In Ritt, Nikolaus / Schendl, Herbert / Dalton-Puffer, Christiane / Kastovsky, Dieter (eds), *Proceedings of the 13th International Conference on English Historical Linguistics*. Bern: Peter Lang.

Smith, Nicholas 2003. Changes in the Modals and Semi-modals of Strong Obligation and Epistemic Necessity in Recent British English. In Facchinetti, Roberta *et al.* (eds), *Modality in Contemporary English*. Berlin: Mouton, 241-267.

Swales, John 1990. *Genre Analysis: English in Academic and Research Settings*. Cambridge: Cambridge University Press.

Taavitsainen, Irma / Pahta, Päivi / Leskinen, Noora / Ratia, Maura / Suhr, Carla 2002. Analysing Scientific Thought-styles: What can Linguistic Research Reveal about the History of Science? In Raumolin-Brunberg, Helena *et al.* (eds), *Variation Past and Present. VARIENG Studies on English for Terttu Nevalainen*. Helsinki: Société Néophilologique, 251-270.

Tiersma, Peter M. 1999. *Legal Language*. Chicago: Univ. of Chicago Press.

Williams, Christopher 2005. *Tradition and Change in Legal English: Verbal Constructions in Prescriptive Texts*. Bern: Peter Lang.

Williams, Christopher forthcoming. Legal English and the 'Modal Revolution'. In Busuttil, Pierre / Salkie, Raphael / van der Auwera, Johan (eds), *Proceedings of the 2nd International Conference on Modality in English*, Pau 2004. Berlin: Mouton.

Notes on Contributors

DAWN ARCHER is a Lecturer of English Language and Linguistics at the University of Central Lancashire. Her main research relates to the discoursal practices of the (historical) English courtroom. Other interests include (historical) pragmatics more generally and also corpus annotation / tool development. Recent publications include *Questions and Answers in the English Courtroom, 1640-1760* (John Benjamins 2005). She is currently compiling a corpus of nineteenth-century English trials. She is also developing a tool (with Paul Rayson) that detects and normalises spelling variants to their modernised forms, so that historical (and dialectal) corpora become more amenable to further annotation and analysis.

NICHOLAS BROWNLEES is Associate Professor of English Language at the University of Florence. His main research interests focus on language and discourse in seventeenth-century English news publications and pamphlets. He is the author of *Corantos and News-books: Language and Discourse in the First English Newspapers* and editor of *News Discourse in Early Modern Britain* (Peter Lang 2006). He is the coordinator of the Florence Early English Newspapers (FEEN) Corpus consisting of news pamphlets of the English Civil War and Inter-regnum. He is currently researching the discourse strategies of Civil War newsbooks.

JOANNA BUGAJ is a Lecturer at Adam Mickiewicz University in Poznań, Poland, and she is currently a Teaching Fellow in the English Language at the University of Leeds. Her main research areas include the history of Scots, the history of English and sociohistorical linguistics and dialectology. She published a monograph on *Middle Scots Inflectional System in the South-West of Scotland* (Peter Lang 2004). In her recent work, she concentrates on discourse-specific features and their link to the issues of standardisation. She is also interested in varieties of English and language policy and planning from a historical perspective.

GABRIELLA DEL LUNGO CAMICIOTTI is Professor of English at the University of Florence. Her key research interests include the pragmatic and discursive analysis of both narrative and academic/ professional texts from a diachronic perspective. Recent publications include various articles on both business and economics discourse of the nineteenth century, on hagiographic narrative in the late Middle Ages, and two books (both co-edited with Elena Tognini Bonelli): *Academic Discourse: New Insights into Evaluation* (Peter Lang 2004) and *Strategies in Academic Discourse* (Benjamins 2005).

MARINA DOSSENA is Professor of English Language at the University of Bergamo (Italy). Her research interests focus on the features and origins of British varieties of English and the history of specialized discourse. Recent publications include *Insights into Late Modern English*, co-edited with Charles Jones (Peter Lang 2003), *Methods and Data in English Historical Dialectology*, co-edited with Roger Lass (Peter Lang 2004), and the monograph *Scotticisms in Grammar and Vocabulary* (Birlinn 2005). She is currently compiling a corpus of nineteenth-century Scottish correspondence (in co-operation with Richard Dury) and is the editor of the *Online Bibliography of Scots and Scottish English*.

HANS LANDQVIST is Associate Professor of Swedish at the University of Göteborg. His main research interests are LSP, especially legal language, lexicology and lexicography, and translation studies. His most recent monograph deals with the development of the Swedish legal lexicon from 1800 to 2000: *Abolition, militärmål, rusdryck och kåkfarare. Studier i ett svenskt juridiskt ordförråd 1800-2000* ['Abolition, military cases, strong drink, and jailbirds. Studies in a Swedish legal lexicon 1800-2000'] (Göteborg 2003). He is currently investigating norms in the translation of popular literature from English into Swedish, and the jargon of professional musicians.

MARJORIE LORVIK is Associate Professor of English Language at Agder University College, Kristiansand, Norway. Her research interest is Scots, and she is currently working on its place in the North Sea language area from a historical point of view. Previous publications include *The Scottis Lass Betrayed? Scots in English Classrooms* (SCCC 1995), an investigation into the role of the secondary school in a revival of Scots.

MARTTI MÄKINEN is a doctoral student at the Research Unit for Variation, Contacts and Change in English, at the Department of English, University of Helsinki. His research interests include corpus linguistics, manuscript and genre studies, and historical linguistics. He has published articles on the medieval English herbal genre, and is now preparing his dissertation: 'Between Herbals *et alia*: A Study of Intertextuality in Medieval English Herbals and Other Contemporary Medicine' He is also one of the compilers of the corpus of Middle English Medical Texts (Benjamins 2005).

PÄIVI PAHTA is Fellow of the Helsinki Collegium for Advanced Studies and a member of the Research Unit for Variation, Contacts and Change in English, both at the University of Helsinki. She has authored and co-edited articles and books on the history of medical and scientific writing, corpus linguistics and language-contact phenomena. Her main publications include *Medieval Embryology in the Vernacular: The Case of* De spermate (1998) and *Medical and Scientific Writing in Late Medieval English* (with Irma Taavitsainen: Cambridge University Press 2004). She is also a co-compiler of the electronic corpora of Middle English Medical Texts (Benjamins 2005) and Early Modern English Medical Texts (in preparation).

ELENA SEOANE is Associate Professor of English in the Department of English at the University of Santiago de Compostela (Spain), and a member of the research project *Variation and Linguistic Change*, directed by Professor Teresa Fanego. In 1996 she received her PhD in English Historical Linguistics from the University of Santiago. Her main areas of specialization are syntactic change in the history of English, historical pragmatics and grammaticalization theory. Her latest publications include a chapter in the *Handbook on the History of English* (Blackwell 2006). More recently she has been interested in scientific English, both in a synchronic and a diachronic perspective, and has written a number of articles on its stylistic changes.

PAOLA TORNAGHI is Associate Professor of English language and linguistics at the Faculty of Economics of the State University of Milano-Bicocca. Her publications include a volume (with M.L. Maggioni) and various articles. Her research interests focus on the phonetics and phonology of Germanic languages, in particular of Old and Middle English and Early Modern English; English vocabulary in both a synchronic and a diachronic perspective; and word-order in Old English prose. She has also studied some of the main aspects of Old English glosses to Latin texts, and Anglo-Saxon lexicography – in particular, the 'Anglo-Saxonici-Latini' dictionaries compiled in the sixteenth and seventeenth centuries.

JUKKA TYRKKÖ is a doctoral student at the Research Unit for Variation, Contacts and Change in English at the University of Helsinki, where he is a member of the Scientific Thought-styles project and participates in the compilation of the Corpus of Early English Medical Writing. He is currently preparing his doctoral dissertation on the discourse analytical and text linguistic aspects of coherence negotiation in narrative hypertexts. His research interests include semiotics, historical pragmatics, lexicography, and lexical semantics.

CHRISTOPHER WILLIAMS is Associate Professor of English at the Faculty of Political Science at Bari University in Italy, and he also teaches English at the Faculty of Law at Foggia University. His main research interests are in the field of tense and aspect in English and also in legal English. His publications include the monographs *Non-Progressive and Progressive Aspect in English* (Schena Editore 2002) and *Tradition and Change in Legal English: Verbal Constructions in Prescriptive Texts* (Peter Lang 2005). He is also co-founder and co-editor of the journal *ESP Across Cultures*.

Linguistic Insights

Studies in Language and Communication

This series aims to promote specialist language studies in the fields of linguistic theory and applied linguistics, by publishing volumes that focus on specific aspects of language use in one or several languages and provide valuable insights into language and communication research. A cross-disciplinary approach is favoured and most European languages are accepted.

The series includes two types of books:

- **Monographs** – featuring in-depth studies on special aspects of language theory, language analysis or language teaching.
- **Collected papers** – assembling papers from workshops, conferences or symposia.

Vol. 1 Maurizio Gotti & Marina Dossena (eds)
Modality in Specialized Texts. Selected Papers of the 1st CERLIS Conference. 421 pp. 2001.
ISBN 3-906767-10-8. US-ISBN 0-8204-5340-4

Vol. 2 Giuseppina Cortese & Philip Riley (eds)
Domain-specific English. Textual Practices across Communities and Classrooms.
420 pp. 2002.
ISBN 3-906768-98-8. US-ISBN 0-8204-5884-8

Vol. 3 Maurizio Gotti, Dorothee Heller & Marina Dossena (eds)
Conflict and Negotiation in Specialized Texts. Selected Papers of the 2nd CERLIS Conference. 470 pp. 2002.
ISBN 3-906769-12-7. US-ISBN 0-8204-5887-2

Editorial address:

Prof. Maurizio Gotti Università di Bergamo, Facoltà di Lingue e Letterature Straniere,
Via Salvecchio 19, 24129 Bergamo, Italy
Fax: 0039 035 2052789, E-Mail: m.gotti@unibg.it

Vol. 4 Maurizio Gotti, Marina Dossena, Richard Dury, Roberta Facchinetti & Maria Lima
Variation in Central Modals. A Repertoire of Forms and Types of Usage in Middle English and Early Modern English.
364 pp. 2002.
ISBN 3-906769-84-4. US-ISBN 0-8204-5898-8

Vol. 5 Stefania Nuccorini (ed.)
Phrases and Phraseology. Data and Descriptions.
187 pp. 2002.
ISBN 3-906770-08-7. US-ISBN 0-8204-5933-X

Vol. 6 Vijay Bhatia, Christopher N. Candlin & Maurizio Gotti (eds)
Legal Discourse in Multilingual and Multicultural Contexts. Arbitration Texts in Europe.
385 pp. 2003.
ISBN 3-906770-85-0. US-ISBN 0-8204-6254-3

Vol. 7 Marina Dossena & Charles Jones (eds)
Insights into Late Modern English.
378 pp. 2003.
ISBN 3-906770-97-4. US-ISBN 0-8204-6258-6

Vol. 8 Maurizio Gotti
Specialized Discourse. Linguistic Features and Changing Conventions.
351 pp. 2003, 2005.
ISBN 3-03910-606-6. US-ISBN 0-8204-7000-7

Vol. 9 Alan Partington, John Morley & Louann Haarman (eds)
Corpora and Discourse.
420 pp. 2004.
ISBN 3-03910-026-2. US-ISBN 0-8204-6262-4

Vol. 10 Martina Möllering
The Acquisition of German Modal Particles. A Corpus-Based Approach.
290 pp. 2004.
ISBN 3-03910-043-2. US-ISBN 0-8204-6273-X

Vol. 11 David Hart (ed.)
English Modality in Context. Diachronic Perspectives.
261 pp. 2003.
ISBN 3-03910-046-7. US-ISBN 0-8204-6852-5

Vol. 12 Wendy Swanson
Modes of Co-reference as an Indicator of Genre.
430 pp. 2003.
ISBN 3-03910-052-1. US-ISBN 0-8204-6855-X

Vol. 13 Gina Poncini
Discursive Strategies in Multicultural Business Meetings.
338 pp. 2004.
ISBN 3-03910-222-2. US-ISBN 0-8204-7003-1

Vol. 14 Christopher N. Candlin & Maurizio Gotti (eds)
Intercultural Aspects of Specialized Communication.
369 pp. 2004.
ISBN 3-03910-352-0. US-ISBN 0-8204-7015-5

Vol. 15 Gabriella Del Lungo Camiciotti & Elena Tognini Bonelli (eds)
Academic Discourse. New Insights into Evaluation.
234 pp. 2004.
ISBN 3-03910-353-9. US-ISBN 0-8204-7016-3

Vol. 16 Marina Dossena & Roger Lass (eds)
Methods and Data in English Historical Dialectology.
405 pp. 2004.
ISBN 3-03910-362-8. US-ISBN 0-8204-7018-X

Vol. 17 Judy Noguchi
The Science Review Article. An Opportune Genre in the Construction of Science.
274 pp. 2006.
ISBN 3-03910-426-8. US-ISBN 0-8204-7034-1

Vol. 18 Giuseppina Cortese & Anna Duszak (eds)
Identity, Community, Discourse.
English in Intercultural Settings.
495 pp. 2005.
ISBN 3-03910-632-5. US-ISBN 0-8204-7163-1

Vol. 19 AnnaTrosborg & Poul Erik Flyvholm Jørgensen (eds)
Business Discourse. Texts and Contexts.
250 pp. 2005.
ISBN 3-03910-606-6. US-ISBN 0-8204-7000-7

Vol. 20 Christopher Williams
Tradition and Change in Legal English. Verbal Constructions in Prescriptive Texts.
216 pp. 2005.
ISBN 3-03910-644-9. US-ISBN 0-8204-7166-6

Vol. 21 Katarzyna Dziubalska-Kołaczyk & Joanna Przedlacka (eds)
English Pronunciation Models: A Changing Scene.
476 pp. 2005.
ISBN 3-03910-662-7. US-ISBN 0-8204-7173-9

Vol. 22 Christián Abello-Contesse, Rubén Chacón-Beltrán, M. Dolores López-Jiménez & M. Mar Torreblanca-López (eds)
Age in L2 Acquisition and Teaching.
214 pp. 2006.
ISBN 3-03910-668-6. US-ISBN 0-8204-7174-7

Vol. 23 Vijay K. Bhatia, Maurizio Gotti, Jan Engberg & Dorothee Heller (eds)
Vagueness in Normative Texts.
474 pp. 2005.
ISBN 3-03910-653-8. US-ISBN 0-8204-7169-0

Vol. 24 Paul Gillaerts & Maurizio Gotti (eds)
Genre Variation in Business Letters.
407 pp. 2005.
ISBN 3-03910-674-0. US-ISBN 0-8204-7552-1

Vol. 25 Ana María Hornero, María José Luzón & Silvia Murillo (eds)
Corpus Linguistics. Applications for the Study of English.
526 pp. 2006.
ISBN 3-03910-675-9 / US-ISBN 0-8204-7554-8

Vol. 26 J. Lachlan Mackenzie & María de los Ángeles Gómez-González (eds)
Studies in Functional Discourse Grammar.
259 pp. 2005.
ISBN 3-03910-696-1 / US-ISBN 0-8204-7558-0

Vol. 27 Debbie G.E. Ho
Classroom Talk. Exploring the Sociocultural Structure of Formal ESL Learning.
254 pages. 2006.
ISBN 3-03910-761-5 / US-ISBN 0-8204-7561-0

Vol. 28 Forthcoming.

Vol. 29 Francesca Bargiela-Chiappini & Maurizio Gotti (eds)
Asian Business Discourse(s).
350 pp. 2005.
ISBN 3-03910-804-2 / US-ISBN 0-8204-7574-2

Vol. 30 Nicholas Brownlees (ed.)
News Discourse in Early Modern Britain.
Selected Papers of CHINED 2004.
300 pp. 2006.
ISBN 3-03910-805-0 / US-ISBN 0-8204-8025-8

Vol. 31 Roberta Facchinetti & Matti Rissanen (eds)
Corpus-based Studies of Diachronic English.
300 pp. 2006.
ISBN 3-03910-851-4 / US-ISBN 0-8204-8040-1

Vol. 32 Marina Dossena & Susan M. Fitzmaurice (eds)
Business and Official Correspondence:
Historical Investigations.
209 pp. 2006.
ISBN 3-03910-880-8 / US-ISBN 0-8204-8352-4

Vol. 33 Forthcoming.

Vol. 34 Forthcoming.

Vol. 35 Vijay K. Bhatia & Maurizio Gotti (eds)
Explorations in Specialized Genres.
316 pp. 2006.
ISBN 3-03910-995-2 / US-ISBN 0-8204-8372-9

Vol. 36 Heribert Picht (ed.)
Modern Approaches to Terminological Theories and Applications.
432 pages. 2006.
ISBN 3-03911-156-6 / US-ISBN 0-8204-8380-X

Vol. 37 Anne Wagner & Sophie Cacciaguidi-Fahy (eds)
Legal Language and the Search for Clarity / Le langage juridique et la quête de clarté. Practice and Tools / Pratiques et instruments.
487 pages. 2006.
ISBN 3-03911-169-8 / US-ISBN 0-8388-5

Vol. 38 Juan Carlos Palmer-Silveira, Miguel F. Ruiz-Garrido & Inmaculada Fortanet-Gómez (eds)
Intercultural and International Business Communication: Theory, Research and Teaching.
343 pages. 2006.
ISBN 3-03910-954-5 / US-ISBN 0-8204-8356-7

Vol. 39 Christiane Dalton-Puffer, Dieter Kastovsky, Nikolaus Ritt & Herbert Schendl (eds)
Syntax, Style and Grammatical Norms: English from 1500–2000.
250 pages. 2006.
ISBN 3-03911-181-7 / US-ISBN 0-8204-8394-X

Vol. 40 Marina Dossena & Irma Taavitsainen (eds)
Diachronic Perspectives on Domain-Specific English.
280 pages. 2006.
ISBN 3-03910-176-0 / US-ISBN 0-8204-8391-5

Vol. 41 John Flowerdew & Maurizio Gotti (eds)
Studies in Specialized Discourse.
293 pages. 2006.
ISBN 3-03911-178-7 / US-ISBN 0-8204-8393-1

Vol. 42 Ken Hyland & Marina Bondi (eds)
Academic Discourse Across Disciplines.
320 pages. 2006.
ISBN 3-03911-183-3 / US-ISBN 0-8204-8396-6

Vol. 43 Paul Gillaerts & Philip Shaw
The Map and the Landscape: Norms and Practices in Genre.
Forthcoming.
ISBN 3-03910-182-5 / US-ISBN 0-8204-8395-4

FSC
www.fsc.org
MIX
Papier | Fördert gute Waldnutzung
FSC® C083411

Kolibri 360 GmbH
Ferdinand-Jühlke-Straße 7
99095 Erfurt, Deutschland
produktsicherheit@kolibri360.de